Will My Cat Eat My Eyeballs?

"Mortician Caitlin Doughty has written the funniest book you'll ever read about what happens to decaying corpses. . . . You'll laugh till you're cat food."　　—Apple Books

"My daughter hasn't put it down. She reads it out loud to me while we giggle."

—Kari Byron, author of *Crash Test Girl* and former host of *Mythbusters*

"A delightful mixture of science and humor. . . . Entertaining."

—*Library Journal*

"A surprisingly heart-warming read."

—Christy Lynch, *BookPage*

"Doughty's answers are as . . . distinctive as the questions. She blends humor with respect for the dead. . . . Her investigations of ritual, custom, law and science are thorough, and she doesn't shy from naming the parts of Grandma's body that might leak after she is gone."　　—Julia Kastner, *Shelf Awareness*

From Here to Eternity

"Doughty chronicles these practices with tenderheartedness, a technician's fascination, and an unsentimental respect for grief."　　—Jill Lepore, *The New Yorker*

"Doughty is a relentlessly curious and chipper tour guide to the underworld, and the weirder things get, the happier she seems. . . . Her dispatches from the dark side [are] doing us all a kindness—offering a picture of what we're in for, even if we'd rather not know."

—Libby Copeland, *New York Times Book Review*

"[Doughty] writes about death with exceptional clarity and style. *From Here to Eternity* manages to be both an extremely funny travelogue and a deeply moving book about what death means to us all." —Dylan Thuras, cofounder of Atlas Obscura

"Doughty writes bluntly about open-air cremations, natural burials and body composting, bringing a little more clarity and a little less mystery to the question, 'What happens to us after we die?'" —Samantha Balaban, NPR, "Our Guide to 2017's Great Reads"

Smoke Gets in Your Eyes

"A trustworthy tour guide through the repulsive and wondrous world of death." —Rachel Lubitz, *Washington Post*

"Demonically funny dispatches."
—Sarah Meyer, O, *The Oprah Magazine*

"Fascinating, funny, and so very necessary, *Smoke Gets in Your Eyes* reveals exactly what's wrong with modern death denial."
—Bess Lovejoy, author of *Rest in Pieces: The Curious Fates of Famous Corpses*

"Think Sloane Crosley meets *Six Feet Under*."
—Kevin Nguyen, *Grantland*

"Morbid and illuminating."

—Stephan Lee, Entertainment Weekly

"*Smoke Gets In Your Eyes* is witty, sharply drawn, and deeply moving. Like a poisonous cocktail, Caitlin Doughty's memoir intoxicates and enchants even as it encourages you to embrace oblivion; she breathes life into death."   —Dodai Stewart

Will My Cat Eat My Eyeballs?

ALSO BY CAITLIN DOUGHTY

From Here to Eternity

Smoke Gets in Your Eyes

Will My Cat Eat My Eyeballs?

And Other Questions about Dead Bodies

Caitlin Doughty

Illustrations by Dianné Ruz

W. W. NORTON & COMPANY

Independent Publishers Since 1923

For information about permission to reproduce selections from
this book, write to Permissions, W. W. Norton & Company, Inc.,
500 Fifth Avenue, New York, NY 10110

For information about special discounts for bulk purchases,
please contact W. W. Norton Special Sales at
specialsales@wwnorton.com or 800-233-4830

Manufacturing by LSC Communications
Book design by Lovedog Studio
Production manager: Lauren Abbate

The Library of Congress has cataloged a previous edition as follows:

Names: Doughty, Caitlin, author. | Ruz, Dianne, author.
Title: Will my cat eat my eyeballs? : big questions from tiny mortals
   about death / Caitlin Doughty, Dianne Ruz.
Description: First Edition. | New York : W. W. Norton & Company,
   2019. | Includes bibliographical references and index.
Identifiers: LCCN 2019020020 | ISBN 9780393652703 (hardcover)
Subjects: LCSH: Death—Anecdotes, facetiae, satire, etc.
Classification: LCC HQ1073 .D68 2019 | DDC 306.9—dc23
LC record available at https://lccn.loc.gov/2019020020

ISBN 978-0-393-35849-0 pbk.

W. W. Norton & Company, Inc.
500 Fifth Avenue, New York, N.Y. 10110
www.wwnorton.com

W. W. Norton & Company Ltd.
15 Carlisle Street, London W1D 3BS

5 6 7 8 9 0

To future corpses of all ages

Contents

Before We Begin

Oh, hey. It's me, Caitlin. You know, the mortician from the internet. Or that death expert from the radio. Or the weird aunt who gave you a box of Froot Loops and a framed photo of Prince for your birthday. (I'm many things to many people.)

When I was a young person, I had a scary encounter with death. But instead of putting me off forever, the experience made me want to learn more. Over the years I've studied medieval history, worked at a crematory, gone to school for embalming, traveled the world to research death customs, and opened a funeral home.

If there's one thing I've learned, it's that death comes for us all. There's no escaping it. Better to look it straight in its eyes. I promise: it's not so bad.

What is this book?

It's pretty simple. I collected some of the most distinctive, delightful questions I've been asked about death, and then I answered them. It's not rocket science, my friends!

(Note: some of it is, in fact, rocket science. See "What would happen to an astronaut body in space?")

Why are people asking you all these death questions?

Well, again, I'm a mortician, and I'm willing to answer strange questions. Plus, I'm obsessed with corpses. Not in a weird way or anything (nervous laughter).

I've given talks all over the United States, Canada, Europe, Australia, and New Zealand on the wonders of death. My favorite part of these events is the Q & A. That's when I get to hear people's deep fascination with decaying bodies, head wounds, bones, embalming, funeral pyres—the works.

All death questions are good death questions, but the most direct and most provocative questions come from kids. (Parents: take note.) Before I started holding death Q & As, I imagined kids would have innocent questions, saintly and pure.

Ha! Nope.

Young people were braver and often more perceptive than the adults. And they weren't shy about guts and gore. They wondered about their dead parakeet's everlasting soul, but really they wanted to know how fast the parakeet was putrefying in the shoebox under the maple tree.

That's why all the questions in this book come from 100 percent ethically sourced, free-range, organic children.

Isn't all this a little morbid?

Here's the deal: It's normal to be curious about death. But as people grow up, they internalize this idea that wondering about death is "morbid" or "weird." They grow scared, and criticize other people's interest in the topic to keep from having to confront death themselves.

This is a problem. Most people in our culture are death illiterate, which makes them even more afraid. If you know what's in a bottle of embalming fluid, or what a coroner does, or the definition of a catacomb, you're already more knowledgeable than the majority of your fellow mortals.

To be fair, death is hard! We love someone and then they die. It feels unfair. Sometimes death can be violent, sudden, and unbearably sad. But it's also reality, and reality doesn't change just because you don't like it.

We can't make death fun, but we can make learning about death fun. Death is science and history, art and literature. It bridges every culture and unites the whole of humanity!

Many people, including me, believe that we can control some of our fears by embracing death, learning about it, and asking as many questions as possible.

In that case, when I die, will my cat eat my eyeballs?

Great question. Let's get started.

Will My Cat Eat My Eyeballs?

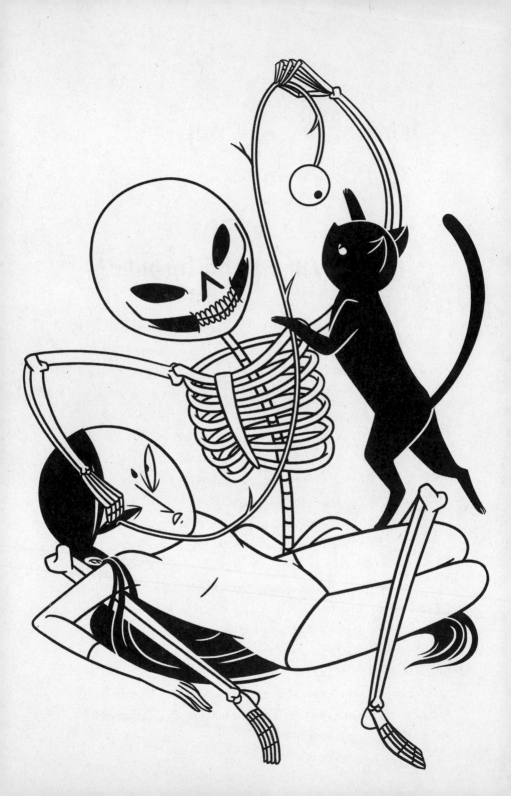

When I die, will my cat eat my eyeballs?

No, your cat won't eat your eyeballs. Not right away, at least.

Don't worry, Snickers McMuffin hasn't been biding his time, glaring at you from behind the couch, waiting for you to take your last breath to be all, "Spartans! Tonight, we dine in hell!"

For hours, even days, after your death, Snickers will expect you to rise from the dead and fill his normal food bowl with his normal food. He won't be diving straight for the human flesh. But a cat has got to eat, and you are the person who feeds him. This is the cat-human compact. Death doesn't free you from performing your contractual obligations. If you have a heart attack in your living room and no one finds you before you miss your coffee date with Sheila next Thursday, a hungry and impatient Snickers McMuffin may abandon his empty food bowl and come check out what your corpse has to offer.

Cats tend to consume human parts that are soft and exposed, like the face and neck, with special focus on the mouth and nose. Don't rule out some chomps on the eyeballs—but Snickers is more likely to go for the softer, easier-access choices. Think: eyelids, lips, or tongue.

"Why would my beloved do that?" you ask. Let's keep in mind that, as much you adore your domesticated meowkins, that sucker is an opportunistic killer that shares 95.6 percent of its DNA with lions. Cats (in the United States alone) slaughter up to 3.7 billion birds every year. If you count other cute little mammals like mice, rabbits, and voles, the death toll might rise to 20 billion. This is an abject massacre—a bloodbath of adorable forest creatures perpetrated by our feline overlords. Mr. Cuddlesworth is a sweetheart, you say? "He watches TV with me!" No, ma'am. Mr. Cuddlesworth is a predator.

The good news (for your dead body) is that some pets with slithery, sinister reputations might not have the capacity (or interest) to eat their owners. Snakes and lizards, for example, won't eat you postmortem—unless you happen to own a full-grown Komodo dragon.

But that's the end of the good news. Your dog will totally eat you. "Oh no!" you say. "Not man's best friend!" Oh yes. Fifi Fluff will attack your corpse without remorse. There are cases where forensics experts first suspect a violent murder has occurred, only to discover that the damage was Ms. Fluff attacking the dead body postmortem.

Your dog might not nip and tear at you because she's starving, however. More likely Fifi Fluff will be attempting to wake you up. Something has happened to her human. She's probably anxious and tense. In this situation, a dog might nibble the lips off her owner, just like you bite your nails or refresh your social media feed. We all have our anxiety busters!

One very sad case involved a woman in her forties who was known to be an alcoholic. Often, when she was intoxicated and unconscious, her red setter would lick her face and bite her legs to try to rouse her. After she died, flesh was found missing from her nose and mouth. The setter had tried to rouse her human again and again, with increasing force, but couldn't wake her.

Forensic case studies—did you know that "forensic veterinarian" is a job?—tend to focus on the destruction patterns of larger dogs: for example, the German shepherd that took out both his owner's eyes, or the husky that ate her owner's toes. But the size of the dog doesn't matter when it comes to postmortem mutilation. Take the story of Rumpelstiltskin the chihuahua. His new owner posted a picture on a message board to show him off, and added some "bonus info" which was that "his [old] owner was dead for a considerable time before anyone noticed and he did eat his human to stay alive." Rumpelstiltskin sounds like a bold little survivalist to me.

Somehow, a dog being anxious and overwhelmed makes us feel better about the whole corpse-eating thing. We develop bonds with our pets. We want them to be upset when we die, not licking their chops. But why do we have that expectation? Our pets eat dead animals, just like humans eat dead animals (okay fine, not you vegetarians). Many wild animals will also scavenge a corpse. Even some of the creatures we think of as the most skilled predators—lions, wolves, bears—will happily chow down if they encounter a dead animal in their territory. Especially if they're starving. Food is food and you're dead. Let them enjoy their meal and go about their lives, now with a slightly macabre pedigree. Viva Rumpelstiltskin!

What would happen to an astronaut body in space?

Two words, many problems: Space. Corpse.

Like the vast reaches of space, the fate of an astronaut corpse is uncharted territory. So far, no individual has died of natural causes in space. There have been eighteen astronaut deaths, but all were caused by a bona fide space disaster. Space shuttle *Columbia* (seven deaths, broken apart due to structural failure), space shuttle *Challenger* (seven deaths, disintegrated during launch), *Soyuz 11* (three deaths, air vent ripped open during descent, and the only deaths to have technically happened in space), *Soyuz 1* (one death, capsule parachute failure during reentry). These were all large-scale calamities, with bodies recovered on Earth in various states of intactness. But we don't know what would happen if an astronaut had a sudden heart attack, or an accident during a space walk, or choked on some of that freeze-dried ice cream on the way to Mars. "Umm, Houston, should we float him over to the maintenance closet or . . . ?"

Before we talk about what would be done with a space corpse, let's lay out what we suspect might happen if death occurred in a place with no gravity and no atmospheric pressure.

Here's a hypothetical situation. An astronaut, let's call her Dr. Lisa, is outside the space station, puttering away on some routine repair. (Do astronauts ever putter? I assume everything they do has a specific, highly technical purpose. But do they ever spacewalk just to make sure everything looks tidy around the ol' station?) All of a sudden, Lisa's puffy white space suit is struck by a tiny meteorite, ripping a sizable hole.

Unlike what you may have seen or read in science fiction, Lisa's eyes won't bulge out of her skull until she finally shatters in a blast of blood and icicles. Nothing so dramatic will occur. But Lisa will have to act quickly after her suit is breached, as she will lose consciousness in nine to eleven seconds. This is a weirdly specific, kind of creepy time frame. Let's call it ten seconds. She has ten seconds to get herself back into a pressurized environment. But such a rapid decompression will likely send her into shock. Death will come to our poor putterer before she even knows what is happening.

Most of the conditions that will kill Lisa come from the lack of air pressure in space. The human body is used to operating under the weight of the Earth's atmosphere, which cradles us at all times like a planet-sized anti-anxiety blanket. From the moment that pressure disappears, the gases in Lisa's body will begin to expand and the liquids will turn into gas. Water in her muscles will convert into vapor, which will collect under Lisa's skin, distending areas of her body to twice their normal size. This will lead to a freaky Violet Beauregarde situation, but will not actually be her main issue in terms of survival. The lack of pressure will also cause nitrogen in her blood to form gas bubbles, causing her enormous pain, similar to what deep-water divers experience when they get the bends. When Dr. Lisa passes out in nine to eleven seconds, it will bring her merciful relief. She will continue floating and bloating, unaware of what is happening.

As we pass the minute and a half mark, Lisa's heart rate and blood pressure will plummet (to the point where her blood may begin to boil). The pressure inside and outside her lungs will be so different that her lungs will be torn, ruptured, and bleed-ing. Without immediate help, Dr. Lisa will asphyxiate, and we'll have a space corpse on our hands. Remember, this is what we think will happen. What little information we have comes from studies done in altitude cham-bers on unfortunate humans and even more unfortunate animals.

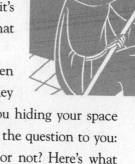

The crew pulls Lisa back inside, but it's too late to save her. RIP Dr. Lisa. Now, what should be done with her body?

Space programs like NASA have been pondering this inevitability, although they won't talk about it publicly. (Why are you hiding your space corpse protocol, NASA?) So, let me pose the question to you: should Lisa's body come back to Earth or not? Here's what would happen, based on what you decide.

Yes, *bring Lisa's body back to Earth*

Decomposition can be slowed down in cold temperatures, so if Lisa is coming back to Earth (and the crew doesn't want the effluents of a decomposing body escaping into the living area of the ship), they need to keep her as cool as possible. On the International Space Station, astronauts keep trash and food waste in the coldest part of the station. This puts the brakes on the bacteria that cause decay, which decreases food rot and helps the astronauts avoid unpleasant smells. So maybe this is where Lisa would hang out until a shuttle returned her to Earth. Keeping fallen space hero Dr. Lisa with the trash is not the best public relations move, but the station has limited room, and the trash area already has a cooling system in place, so it makes logistical sense to put her there.

Yes, Lisa's body should come back, but not right away

What if Dr. Lisa dies of a heart attack on a long journey to Mars? In 2005, NASA collaborated with a small Swedish company called Promessa on a design prototype for a system that would process and contain space corpses. The prototype was called the Body Back. ("I'm bringing body back, returning corpses but they're not intact.")*

If Lisa's crew had a Body Back system on board, here's how it would work. Her body would be placed in an airtight bag made of GoreTex and thrust into the shuttle's airlock. In the airlock, the temperature of space (−270°C) would freeze Lisa's body. After about an hour, a robotic arm would bring the bag back inside the shuttle and vibrate for fifteen minutes, shattering frozen Lisa into chunks. The chunks would be dehydrated, leaving about fifty pounds of dried Lisa-powder in the Body Back. In theory, you could store Lisa in her powdered form for years before returning her to Earth and presenting her to her family just like you would a very heavy urn of cremated remains.

* Kids, this is a Justin Timberlake reference, you're fine not knowing who that is.

Nope, Lisa should stay in space

Who says Lisa's body needs to come back to Earth at all? People are already paying $12,000 or more to have tiny, symbolic portions of their cremated remains or DNA launched into Earth's orbit, to the surface of the moon, or out into deep space. How psyched do you think space nerds would be if they had the chance to float their whole dead body through space?

After all, burial at sea has always been a respectful way to put sailors and explorers to rest, plopped over the side of the ship into the waves below. We continue the practice these days despite advances in onboard refrigeration and preservation technology. So, while we do have the technology to build robot arms to shatter and freeze-dry space corpses, perhaps we could employ the simpler option of wrapping Dr. Lisa in a body bag, space-walking her past the solar array, and letting her float away?

Space seems vast and uncontrolled. We like to imagine that Dr. Lisa will drift forever into the void (like George Clooney in that space movie I watched on the plane that one time), but more likely she would just follow the same orbit as the shuttle. This would, perversely, turn her into a form of space trash. The United Nations has regulations against littering in space. But I doubt anyone would apply those regulations to Dr. Lisa. Again, no one wants to call our noble Lisa trash!

Humans have struggled with this challenge before, with grim results. There are only a few climbable routes to climb to the top of Mount Everest's 29,029-foot peak. If you die at that altitude (which almost three hundred people have done), it is dangerous for the living to attempt to bring your body down for burial or cremation. Today, dead bodies litter the climbing paths, and each year new climbers have to step over the puffy orange snowsuits and skeletonized faces of fellow climbers. This same thing could happen in space, where shuttles to Mars have to pass the orbiting corpse every trip. "Oh geez, there goes Lisa again."

It's possible the gravity of a planet could eventually pull Lisa in. If that happens, Lisa would get a free cremation in the atmosphere. Friction from the atmospheric gas would super-heat her body's tissues, incinerating her. There's the smallest of small possibilities that if Lisa's body was sent out into space in a small, self-propelled craft like an escape pod, which then departed our solar system, traveled across the empty expanse to some exoplanet, survived its descent through whatever atmosphere might exist there, and cracked open on impact, Lisa's microbes and bacterial spores could create life on a new planet. Good for Lisa! How do we know that alien Lisa wasn't how life on Earth started, huh? Maybe the "primordial goo" from which Earth's first living creatures emerged was just Lisa decomposition? Thanks, Dr. Lisa.

Can I keep my parents' skulls after they die?

Ah yes, the old "Can I keep my relatives' skulls?" question. You'd be surprised (or maybe you wouldn't be surprised) how often I am asked this question.

Wait. First of all, what are you going to *do* with their skulls, exactly? Set them on your mantel? Transgressive Christmas tree topper? Whatever your plans are, remember, real skulls aren't kitschy Halloween decorations; they belonged to a human being. But assuming your intentions are good, you're looking at three major hurdles to clear before Dad's brainpan can hold jellybeans on your coffee table: paperwork, legal control, and skeletonization.

First, let's talk paperwork. It is extremely difficult to get legal permission to display a relative's skeleton. In theory, people get to decide what happens to their bodies after death. So, *in theory* your parents could create a written, signed, dated document explicitly stating that they want you to have their skull after they die. It would be similar to the document a person signs if they want to donate their body for scientific research.

Tell you what's not going to work: marching on over to your local funeral home and saying, "Greetings! That's my mom's corpse over there. Could you just pop off her head and de-flesh her skull? That would be great. Thanks!" Your average

funeral home (really, any funeral home) is not set up to handle such a request, legally or practically. As a funeral director, I honestly have no idea what equipment a proper decapitation requires. The subsequent de-fleshing is far beyond me. I assume it involves boiling and/or dermestid beetles, but that's not in the mortuary school curriculum.

(My editor wrote this note here: "To be fair, you *do* actually know a thing or two about de-fleshing." Okay, true. I've never done it on a human, but I am an amateur dermestid beetle enthusiast. The beetles are incredible creatures, used in museums and forensic labs to delicately eat the dead flesh off a skeleton without destroying the bones themselves. Dermestids are happy to wade into a gruesome, sticky mass of decaying flesh and delicately clean around even the tiniest of bones. But don't worry about visiting a museum and accidentally falling into a vat of dermestids: despite being "flesh-eating" beetles, they aren't interested in the living.)

Back to Mom's head. Even if I could remove it, my funeral home could not legally hand over the decapitated head because of a topic that will come up multiple times in this book: abuse of corpse laws. Abuse of corpse laws vary from place to place, and can sometimes seem a little arbitrary. For example, the law in Kentucky says you're committing corpse abuse if you treat a dead body in a way that "would outrage ordinary family sensibilities." But what is an "ordinary family?" Maybe in your "ordinary family" Dad was a scientist who always promised that when he died he would leave you both his collection of Bunsen burners and his skull. There are no ordinary families.

Abuse of corpse laws exist for a reason, however. They protect people's bodies from being mistreated (ahem, necrophilia). They also prevent a corpse from being snatched from the morgue and used for research or public exhibition without the dead person's consent. You'd be surprised how often this has happened throughout history. Medical professionals have stolen corpses and even dug up fresh graves to get bodies for dissection and research. Then there are cases like that of Julia Pastrana, the nineteenth-century Mexican woman with a condition called hypertrichosis that caused hair to grow all over her face and body. After she died, her embalmed and taxi-  dermied corpse was taken on world tour by her awful husband. He saw there was money to be made by displaying Julia in freak shows. Julia had ceased to be regarded as human; her corpse had become a possession.

Because of abuse of corpse laws, nobody's dead body can be *claimed* as property. "Finders keepers" doesn't apply here. But unfortunately, those same abuse of corpse laws prevent you from plopping Mom's skull on your bookcase.

"Wait, I've seen human skulls on people's bookcases before! How did they get them?" In the United States, there is no federal law that stops the ownership, buying, or selling of human remains. Well, except if the remains are Native American. In that case, you're out of luck (and rightfully so). But otherwise, whether you're able to sell or own human remains is decided by each individual state. At least thirty-eight states have laws that should prevent the sale of human remains, but in reality the laws are vague, confusing, and enforced at random.

In one seven-month period in 2012–13, there were 454 human skulls listed on eBay, with an average opening bid of $648.63 (eBay subsequently banned the practice). Many skulls for private sale have questionable origins, sourced from the thriving bone trades in India and China. The bones are obtained from people who couldn't afford cremation or burial—not exactly ethical sourcing. These plucky bone sellers will tell you that it's not human *remains* they sell, but human *bones*. Most state laws prohibit the sale of "remains," but bones are totally legal and in keeping with the law, they will say.

(Note: they are selling remains.)

So, to be clear: you can't own your mom's corpse, but if you are willing to engage in some suspect internet commerce, a random Indian person's femur might make its way into your house.

Even if you exploit fuzzy legal arguments in your quest to get your hands on Dad's skull, you're still going to run into a big problem: there is currently no way in the United States to skeletonize human remains for private ownership. For the most part, skeletonization only happens when a body is donated to scientific research. *Even this* isn't explicitly legal (authorities just tend to look the other way for museums and universities). But under no circumstances can you just skeletonize your dad and plop his head among the decorative gourds in the Thanksgiving centerpiece.

I spoke to my friend Tanya Marsh, a law professor specializing in human remains law. She's the expert on this stuff. If there is *any* legal wiggle room that might allow a person to get Dad's head liberated from its fleshy shell, Tanya would know how to find it.

ME: People ask me about this all the time, there has to be a way.

TANYA: I will argue with you all day long that it isn't legal

in *any state* in the United States to reduce a human
head to a skull.

ME: But if it was donated to science and then donated to
the famil—

TANYA: No.

In every state, funeral homes use something called a burial
and transit permit, which tells the state what is going to be done
with the dead person's body. The options are usually burial, cremation, or donation to science. That's it: three simple things.
There is no "cut off the head, de-flesh it, preserve the skull, and
then cremate the rest of the body" option. Nothing even close.

Tanya read to me the fine print from one state law:

> . . . *every person who deposits or disposes of any human
> remains in any place, except in a cemetery, is guilty of a
> misdemeanor.*

In other words, Dad's skull is supposed to be in the cemetery,
and you're committing a crime if you put it anywhere that's *not*
a cemetery, such as your garden.

To offer you a ray of hope, laws are changing as I write this.
Right now, owning human bones (your mom's or someone else's)
is a big, murky, gray area. Maybe someday laws will change in
your favor, and Your Momz Skull, LLC, will specialize in legally
de-fleshing parental skeletons.

If that's what you (*and* your parents!) want, I hope it can
happen for you. If all else fails, cremate them and press their
ashes into a diamond or a vinyl record. Kids, a vinyl record
is . . . never mind.

Will my body sit up or speak on its own after I die?

Come closer, deathlings. I'm not sure I should be telling you this; the morticians' secret council will be very angry with me. But one night, I was working late in the funeral home, alone. In the body preparation room, stretched on the table under a white sheet, was a dead man in his forties. As I reached to turn off the light, a long horrifying moan came from the body, and the man sat straight up, like Dracula rising from his coffin . . .

Okay, that never happened. I made it up. (Not the part about working late—every funeral employee has to work late.) But this story, or something like this story, is everyone's favorite morgue or funeral home twisted tale. It usually comes from a source like "my husband's cousin's nephew" who worked at a funeral home in the 1980s and once saw a body sit up. You'll find the stories on message boards and articles with titles like "Creepy Stories Funeral Directors Don't Want You to Know."

But what are the facts on postmortem movement?

Your body is not going to sit bolt upright on its own corpsey power. This is not a horror movie, folks. Dead bodies aren't going to scream, sit up, or grab your hair and pull you down to hell (although I'll admit I had some of these exact unfounded fears when I started working at a funeral home).

However, just because your corpse isn't showing off with these grand, "look at me!" moves, that doesn't mean there aren't a range of twitches, spasms, and groans a dead body

might make. You're thinking, a twitching dead body is still pretty freaky! I hear you. But there are simple biological reasons for how and why such a thing could happen.

Right after a person dies, their nervous system may still be firing, which can cause small spasms and twitches of the body. These spasms usually happen in the first few minutes after death, but sometimes they're observed up to twelve hours later. As for noises, when a recently dead body is moved, air can be pushed out of the windpipe, creating an eerie moan. Most nurses have experienced some of these things, so after a person has been declared dead, their response to a twitch, movement, or moan tends to be calmer, not "Dear God, it's alive, it's ALIIIVVE!"

Your body might also make noises that have nothing to do with the dying nervous system. After you die, your gut is party central, with billions of bacteria eating away at your intestines before moving on to your liver, your heart, your brain. But, with all that feasting comes waste. Those billions of bacteria produce gases like methane and ammonia, which bloat your stomach. That bloat means internal pressure, and if the pressure builds up enough, your body can purge, releasing vile-smelling liquid or air. When a body purges, it may make a creepy whooshing sound. Worry not, this isn't the horrible ghost wails of the dead, it's . . . bacteria farts.

Groaning corpses have fascinated people for centuries. Before we knew about bacteria farts and the nervous system, and before we had clearer, scientific definitions of death, people were terrified of being buried alive. Twitches and moans made the dead person seem not quite so dead.

In Germany, in the late 1700s, there were physicians who believed that the only way to tell if someone was truly dead was to wait for the person to start rotting—bloating, smelling, the whole works. This belief led to the creation of the Leichenhaus, a "waiting mortuary," where dead bodies would hang out in a fire-heated room (heat encourages decomposition) until no one could dispute that the dead person was 100 percent dead. The rooms would be watched by a young male attendant in case anyone should moan, sit up, request the bathroom, whatever. They often attached bells to the corpses, which would ring if the body moved, and alert the attendant. In practice, what this amounted to was a young man sitting in a silent room filled with hideously stinky corpses.

One of these waiting mortuaries, in the city of Munich, charged a fee for visitors to walk among the corpses. They created a "look out, corpse is alive!" alarm system by tying strings to the fingers and toes of the dead body. The strings were attached to a harmonium (an organ that makes sound when air blows through it). Any movement was supposed to trigger the instrument and alert the attendant if a corpse was moving around. This worked, but unfortunately the "movement" was just the swelling and bursting of the decomposing body. In the night the attendant would wake up to an empty room filled with a creepy, tuneless melody.

By the late 1800s, most of the waiting mortuaries had ceased operations. A Dr. von Steudel said that a million bodies had passed through the mortuaries, and not a single one had woken up.

The answer here is yes, dead bodies can move by themselves, but the movements are small, and caused by science! Not ghosts. Or demons. Or zombies. Just be glad you're not an attendant at the Leichenhaus.

We buried my dog in the backyard, what would happen if we dug him up now?

There are many reasons you might want to resurrect your dog from his spot under the maple tree. Unlike with human burials, there aren't laws that prevent you from peeking in on your pup to see how his decomposition is going. (Note: In human cemeteries unlawful exhumation, or digging up bodies without a permit, is considered grave desecration. I don't want to hear you claiming, "Caitlin said to check on how Grandma is coming along.")

The most common reason people dig up their pets is because they are moving. They can't bear to leave Growler the Pekingese behind, and don't want some new family who didn't even know Growler building a swimming pool and sending his bones away in a dump truck. But they might also be feeling squeamish about what Growler looks like eight months after burial. Enter companies that will come to your house, dig up Growler, and have him cremated and brought back to you. Now residing in his bone-shaped urn, Growler is ready to travel to his new home.

As for what Growler will look like when he's dug up, there are so many factors it's almost impossible to answer hypothetically. One pet exhumation specialist from Australia offered this general rule of thumb: "When you exhume pets that are fifteen years old, you are looking for bones, whereas if you have ones that are one to three years old, they're a bit more intact and smelly." But this timeline depends on many other factors. How long has he been dead? Was he put in a Growler-sized coffin or straight into the soil? Where do you live: tropical rainforest, desert, grassy suburb? I need more information!

How deep was Growler buried? He'll decompose more slowly if you dug way down, many meters under that maple tree. The deeper he is buried, the farther away he is from oxygen, microbes, and other things that speed up the decomposition process.

What kind of soil was Growler buried in? This might be the biggest factor in what Growler looks like now. All soils aren't simply "whatever . . . dirt, right?" Soils are as different as the colors of the rainbow.

Egypt, for example, is known to have sandy soil, which can preserve bone very well. It's also known for being very hot. That combination, dry and hot, could have dehydrated Growler and mummified him. In the scorching sand, his skin would have dried so quickly and thoroughly that not even bugs could chomp through it. Animal mummies are more common than you might think. In 2016, a zoo in the Gaza Strip had to be abandoned due to war and the Israeli blockade. As the animals died one by one, they mummified in the dry, hot air. Pictures from inside the ghost zoo show eerily preserved lions, tigers, hyenas, monkeys, and crocodiles.

Hundreds of years ago, across Europe, people afraid of witchcraft would seal cats inside the walls of their homes, believing they would ward off supernatural threats. Builders and contractors have been finding random cats in European walls for years. A shop owner in England had a customer bring in a box containing a mummified cat and a mummified rat, both over three hundred years old. The customer had found them in the walls of a Welsh cottage and wanted to sell them. This is all to say that if conditions are right, you may have a mummified Growler on your hands.

Notably, there was a dog named Stuckie found in Georgia in the 1980s. Stuckie was likely a hunting dog that ran up the inside of a hollow tree after a squirrel. As Stuckie climbed, the trunk became narrower and (you see where this is going) Stuckie got stuck. Loggers found his mummified body in the tree years later, teeth bared, eye sockets empty, toenails still intact. They could see all of Stuckie's bones showing through his thin mummified skin and fur. Normally he would have decomposed quickly in the Georgia woods, but since no creatures could get access to eat him, and the tree bark and tannins sucked moisture from his skin, Stuckie became immortal.

Stuckie's case is rare. You might hope to find immortal Growler buried in the backyard, but you're more likely to find no Growler at all. The ideal gardening soil is loamy: silt, sand, and clay mixed together. Loamy soil is also ideal animal decomposing soil. If Growler was buried in summer, when temperatures are high, and close to the surface, where the soil has just the right amount of moisture, oxygen, and microbes, the loam might have decomposed all of Growler's soft gooey tissue, skin, and organs, and even his bones!

The location and depth of soil you choose will determine your dog's postmortem fate (or your gerbil's, or ferret's, or turtle's). Do you want him to become part of the garden? If so, bury him close to the surface or in rich soil, where he has the best chance of decomposing fast and completely. If you want him to stay around longer, wrap him in plastic and place him in a sealed box, buried deep down in the soil. Although if you really want Growler to stick around for the long haul, may I recommend taxidermy?

Can I preserve my dead body in amber like a prehistoric insect?

This is a fantastic question. You, young person, are a pint-sized death revolutionary. Everyone should be on the lookout for new possibilities for our future corpses. Let's hang out and brainstorm ideas sometime.

I think a dead body encased in amber would be cool as heck. You've probably seen pictures of perfect-looking ancient insects trapped in a smooth orange casing. The insects are a delivery from another time—a tree resin time machine. Let's talk about how they got stuck in there in the first place. Trees produce resin, that gummy, sticky substance oozing from the bark that's almost impossible to get off your hands even after you wash them seven times. The trees use that resin as protection against various pests and animals that might harm them. Say it's 99 million years ago, and an ancient ant is crawling up a tree and becomes stuck in the resin. The tree's trap has worked; the ant is done for. Soon, more resin covers the poor creature and solidifies. Normally, this resin will be disintegrated by wind, rain, sunlight, or bacteria over time—crumbling along with Monsieur Ant inside. But every once in a while, the resin is protected and preserved so that over many millions of years it fossilizes into amber.

Here is a short, but awesome, list of things that have been found preserved in amber: a roughly 20-million-year-old male scorpion dug up by a farmer in Mexico, a roughly 75-million-year-old set of dinosaur feathers found in Canada, a roughly 17-million-year-old group of anole lizards found in the Dominican Republic, and a roughly 100-million-year-old insect (now extinct) with a triangular head that could turn around 180 degrees— something no modern insect is able to do. There's even a chunk of amber that holds a roughly 100-million-year-old spider paused in mid-attack on a wasp.

All these creatures, long ago, were trapped and preserved in resin. So the question is, why not you? When you die (no need to trap you alive, that's a little grisly; already dead is fine), theoretically we could encase you in tree resin. Maybe, like the spider-wasp fight, we can pose you fighting a panther or something. Then we would place you and the panther (surrounded by resin) into a climate-controlled room, putting you through a series of chemical changes with heat and pressure. If all goes well, a hop, skip, and several million years later the resin will turn to amber. At least, we think it will take several million years; there is no firm answer as to just how long the process of turning resin into amber takes. At this point, some future sentient creature might find you and go, "Whoa, check out this gnarly human trapped in amber." Maybe the creature will use you as a paperweight on their desk or something.

Okay, so you're a human, preserved in amber. But you should know that with the science available right now, there's something they won't be able to do with your fossilized body: clone you. I bring this up because I have a suspicion that you asked this whole trapped-in-amber question because you have some secret *Jurassic Park* dream of "life finding a way." Of DNA being extracted from your amber casing, cloned, and becoming a version of You 2.0.

The idea behind *Jurassic Park*, before it became a book and then a massive movie franchise, began as a thought experiment by some scientists in the 1980s. They looked at an ancient mosquito trapped in amber and wondered, "What if one of those mosquitoes had feasted on the blood of a T. rex right before it died? The mosquito ate, landed on a tree to relax, got caught in the tree resin, and eventually became preserved in amber? If we can suck out that ancient dino blood, maybe we can get its genetic code and use it to bring that T. rex back." I admit this is a neat prospect. And in some ways, amber is fantastic for preserving dead organic material. For one, amber is very, very dry. Dry environments (like deserts) are ideal for preservation. So why wouldn't they be able to get DNA out of those flawlessly preserved creatures in amber?

Scientists now pretty much agree that getting useful DNA from animals in amber is not possible. DNA just disintegrates too quickly. Oxygen levels change, temperatures change, moisture levels change, all of which cause the puzzle pieces that make up your genetic code to fall apart. It's a mess. Even if they did get some of your material extracted, they'd likely have to fill in the gaps with . . . someone, or something, else. For example, scientists at Harvard University are taking genes from extinct woolly mammoths and attempting to "cut and paste" their DNA into elephant cells. If it works, the resulting creature will not be a mammoth but some kind of mammoth-elephant hybrid. Maybe you can be spliced together with the panther you're fighting. Hybrid panther humans of the future! (This is made up, it's not going to happen—don't listen to me, I'm just a mortician.)

You need to decide what's more important to you here. Are you hoping to look good for potentially millions of years, and be the ultimate decorative piece? In that case, encasing you in resin might be a good choice. But if you want to preserve your DNA so that you could perhaps be cloned in the distant future, you might want to consider something else: cryopreservation. When you die, we can flash freeze your cells in liquid nitrogen, way down in the negative degrees Fahrenheit. Scientists have cloned both mice and bulls from cells in a deep freeze.

Maybe what you're looking for is less *Jurassic Park*, more *Star Wars*. Remember when Han Solo was put into a "carbon freeze," a gas that freezes into a solid state? That's not very convincing science either, but it brings you closer to your goal of freezing your cells. There's no evidence that freezing your entire body can bring you, the individual, back to life in the future. But preserving your cells to clone? Just maybe. On another note: our highest-grossing movies seem to involve a lot of sophisticated body preservation technology. Coincidence? I don't think so. The public loves fancy corpse tech. (*Frozen* never really went there, but I have a feeling Elsa has some good cryopreservation techniques up her sleeve.)

So, you may never be cloned. But unlike the dinosaurs (or the quagga, or the woolly mammoth, or the passenger pigeon), humans aren't likely to go extinct anytime soon. There are 7.6 billion of us on Earth—and the number is growing. The conversation in the next fifty years is more likely to be whether it is the responsibility of us humans to bring back animals that we have pushed to extinction, or the brink of extinction. But maybe the conversation a million years from now will be about whether to bring the human race back, and the lucky preserved human may just be you!

Why do we turn colors when we die?

Dead bodies can be a colorful kaleidoscope of activity. That's one of my favorite parts about them. Maybe you are dead—"you," meaning Jessica or Maria or Jeff—but that doesn't mean that life isn't still happening inside your flesh case. Blood, bacteria, fluids: they're reacting, changing, and adapting now that their host is dead. And those changes mean . . . *colors*.

The first colors that appear after death have to do with blood. When a person is alive, blood pumps through their body. Take a look at your fingernails right now. If they're pink, that means your heart is pumping blood. Congratulations, you're alive! Hopefully you don't need a manicure. My nails are a horrific mess right now. That is neither here nor there, so . . . moving on.

In the first hours after death occurs, a dead person will look paler than before, especially in places like their lips and fingernails. They lose their healthy pinkish color and start to turn colorless and waxy, because the blood that once ran right under the surface of the skin has started to succumb to gravity. When you think of a ghastly pale corpse, it's a phenomenon as dull as blood loss in surface tissues.

Around this time, you'll also see a color change in the person's eyeballs. Corpses will need your help in closing their eyes. In my funeral home, we recommend families do this fairly soon after death. In as little as half an hour, the iris and pupil cloud over and turn milky because the fluid under the cornea has stagnated, like a creepy little bog. If this reminds you of a zombie, I recommend you shut the person's eyes. This will make it look more as if the body is sleeping and less "Dad's lifeless clouded-over eyes boring into your soul."

Once the blood starts to settle, you're going to see more dramatic color changes. When you're alive, your blood is made of different components mixed together. But when the blood stops moving, the heavier red blood cells fall slowly out of the mix, like sugar settling to the bottom of a glass of water.

This leads to the first solid, visible sign of death, livor mortis. Livor mortis is the pooling of blood in lower areas of the corpse, usually a person's back. (Again, thanks, gravity.) The pools tend to be purple in color. In Latin, the phrase means "the bluish color of death."

Remember, when we're talking about the "discoloration" of a body after death, we have to remember what color the living person was to start with. Discoloration is more dramatic and obvious on lighter-colored skin. But worry not, these postmortem color changes (like decomposition) come to us all.

Interestingly, livor mortis can be useful to forensic examiners determining how and where someone died. The patches of color, and how intensely purple they are, make a difference. For instance, if the livor mortis is all over the front of the body, that means the corpse has been lying face-down for several hours, giving the blood time to pool there.

However, livor mortis patches *won't* be found on the parts

of the body pressed up against something—the floor, for example—because the pressure means the teeny tiny vessels near the body's surface can't fill up with blood. This is yet another way investigators can tell if a body has been lying in a certain position, or on top of something.

And wait, there's more. What if livor mortis is a different color? If the livor mortis is bright cherry red, that might mean the person died in the cold, or by inhaling carbon monoxide (maybe smoke from a fire). If the livor mortis is deep purple or pink, that might mean the person suffocated, or died of heart failure. Finally, if a person has lost a lot of blood, you might not find any livor mortis at all.

Livor mortis is the first color change you'll see in a dead body during the first several hours. But there's a whole new fabulous bouquet of colors waiting to blossom about a day and a half after death.

Welcome to putrefaction. This is when the famous green color of death comes into its own. It's more of a greenish-brown, actually. With some turquoise. You could call this color "putrid," and you'd be totally correct. The green-purple-turquoise blossoms of putrefaction are caused by bacteria. Remember when I said that even after you die there are still fun things happening inside your flesh case? Well, bacteria are the most important guests at the party. Gut bacteria go wild, digesting you from the inside.

The green colors appear first in the lower abdomen. That's the bacteria from the colon breaking free and starting to take over. They are liquefying the cells of the organs, which means fluids are sloshing free. The stomach swells as gas starts to accumulate from the bacteria's "digestive action" (i.e., bacteria farts).

As the bacteria multiply and spread, so does the green discoloration, eventually ripening to a darker green or black.

Decomposition isn't only about bacteria. Another decomposition process is called autolysis. Autolysis happens when enzymes begin destroying the body's cells from the inside. This  destruction process has been quietly occurring all along—since a few minutes after the person died.

The body is now on a complicated journey, carried along by autolysis and putrefying bacteria. New patterns of color arise. You will start to see signs of a venous patterning, or marbling, of blood vessels near the surface of the skin. This is the classic "purple vein" effect that movie makeup people use to show somebody has been infected by a zombie virus. In a corpse, this marbling is the visible sign of blood vessels decaying and hemoglobin separating from the blood. The hemoglobin stains the skin, producing delicate color schemes in shades of red, dark purple, green, and black. The hemoglobin ring breaks down into bilirubin (turning you yellow) and biliverdin (turning you green).

This technicolor show is happening alongside all the other visible effects of putrefaction, like swelling, "purging," and blistering or peeling of the skin. The color will change so profoundly that you will no longer recognize the person or be able to tell the age or complexion they were in life.

How come you don't usually see bodies in extreme states of decomposition, except in zombie or horror movies? Well, in the twenty-first century, bodies aren't usually allowed to decay to this point. Because you almost never see bodies decomposing in real time, most people seem to believe dead bodies immediately bloat, swell, and turn colors. Not true; it takes days. At a funeral home, the body will either be embalmed (a chemical process that slows decomposition), or put in a refrigeration unit (the cold air slows decomposition). A body will then swiftly be buried or cremated, so a family never comes face to face with the reality of decay. No wonder you are confused about the timeline of decomposition; you are likely to go your whole life without seeing a fully decomposed body! You'll miss the beautiful colors, but considering you'd have to, I don't know, stumble upon a dead body in the woods to see them, maybe it's for the best.

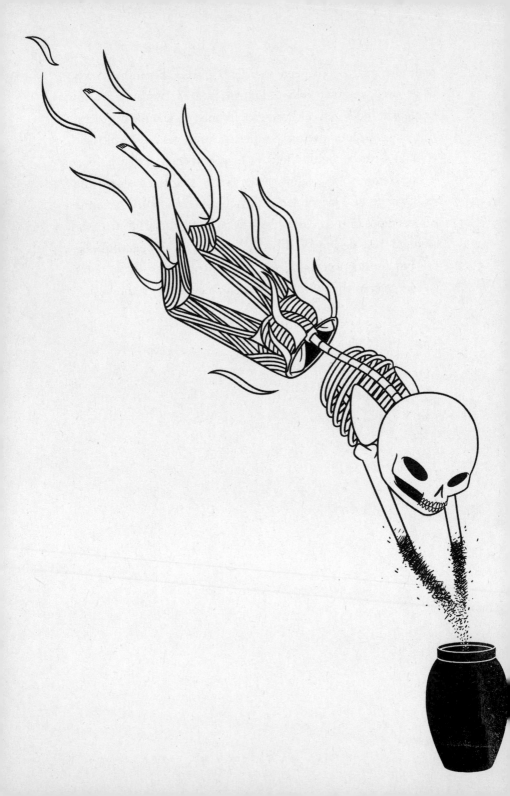

How does a whole adult fit in a tiny box after cremation?

It feels weird when a funeral director hands you a silver urn with doves and roses on it, about the size of a coffee can, and says, "Here's your grandma!" Um, Grandma was a lot bigger than that, thanks very much. It's even stranger when a funeral director hands you the exact same doves-and-roses urn and says, "Here's your neighbor Doug!" Wait a second, Doug was six feet four and 340 pounds—how can he fit in the same urn that Grandma fits in? This cremation thing is a scam!

No, it's not a scam. There's a good reason that people are (mostly) the same size after a cremation.

You know when you're nervous about giving a big speech to a group and they tell you to imagine the audience naked? Here's another fun exercise: imagine the audience as skeletons. Strip away all the skin, fat, and organs, because underneath it all, everyone's skeleton is sort of the same. Some folks are taller, of course, some bones are thicker, some people have only one arm—but for the most part, a skeleton is a skeleton. And whether you're holding an urn containing your grandma or your neighbor Doug, it's a ground-up skeleton in there.

Here's how the cremation process works. When the door to the cremation machine opens, a whole human slides in. They've probably been in refrigerated storage for a few days to a week, but overall things haven't changed that much. They may even be wearing the same clothes they died in. But once the machine door closes and the 1500-plus-degree flames start their work, the body immediately starts to transform.

In the first ten minutes of the cremation, the flames attack the body's soft tissue—all the squishy parts, if you will. Muscles, skin, organs, and fat sizzle, shrink, and evaporate. The bones of the skull and ribs start to emerge. The top of the skull pops off and the blackened brain gets zapped away by the flames. The human body is roughly 60 percent water, and that H_2O—along with other body fluids—evaporates right up the machine's chimney. It takes just a little over an hour for all the organic material in the human body to disintegrate and vaporize.

What are we left with at the end of a cremation? Bones. Hot bones. We call this pulverized mess of molten bones "cremated remains" or, more commonly, ashes. (Funeral directors like to call them "cremated remains" because it sounds fancier and more official, but "ashes" is fine.)

This isn't a full human skeleton, mind you. Remember, the organic material in our bones burns during the cremation. What's left behind in the cremated remains is a thrilling combo of calcium phosphates, carbonates, and minerals and salts. They are totally sterile, which means you could roll around in them like a snowdrift or a sandbox and be perfectly safe. I'm not recommending that, just saying that you could. They don't have DNA left in them, either. It's basically impossible to tell Grandma's bones from neighbor Doug's bones just by looking at them, which is why cremation was long considered the best way to cover up a crime. (Nowadays, if there is any foul play suspected in a death, cremation can't take place until a full investigation is completed.)

After they've cooled down, the bone fragments are swept out of the cremation machine. Any big pieces of metal are removed (Did Grandma have a hip implant? We'll find out when we cremate her!) and the bones are ground down into ashes. The crematory operator pours this light gray powder into an urn, which is given back to the family to scatter, bury, turn into a diamond, shoot into space, make into a painting, or use as tattoo ink.

But what about a person who weighs, say, 450 pounds? Surely those ashes will be heavier. Nope. Much of that weight is fat. Underneath, remember, their skeleton is pretty much identical to everyone else's. Since fat falls into the category of organic material, it will burn up during the cremation process. Cremations for very heavy people can take longer, sometimes over two hours longer. That gives the fat enough time to burn away. But at the end of the process, you can't tell who went in the machine a 450-pound person and who went in a 110-pound person. The flames are the great equalizer.

It's more height than weight that determines how much ash is in that doves-and-roses urn. Women tend to be shorter—less bone—so their ashes usually weigh around four pounds. Men tend to be taller, and have ashes that weigh around six pounds. I'm a woman who is over six feet tall, so I'm hoping to have some pretty heavy ashes if I'm cremated. (I'd rather be eaten by wild animals, but that's a whole other question.) My uncle, who died a few years ago, was six feet five inches tall. His were some of the heaviest ashes I've ever held.

Forget about what you look like on the outside; it's the weight on the inside (your skeleton) that counts. In the end, Grandma and Doug both fit in that tiny urn because all the organic material, including skin, tissue, organs, and fat has vaporized into the air, leaving their brittle bones behind.

If Grandma and neighbor Doug have identical cremated remains, and there is no remaining DNA, is there any difference between the two urns at all? It can feel like there is nothing special about Grandma's ashes, nothing left of her special "Grandma-ness." Not true! There are differences, even if we can't see them. Maybe Grandma was a vegetarian who took multivitamins. Maybe Doug lived near a factory for most of his life. These factors affect what trace elements are found in the ashes.

Grandma's ashes may look and feel similar to Doug's, but Grandma is still Grandma. Which means you'll for sure be swapping that doves-and-roses urn the funeral home gave you for Grandma's custom Harley Davidson urn. She was that kind of gal.

Will I poop
when I die?

You might poop when you die. Fun, right? I enjoy pooping in my day-to-day life, so it's comforting to think this activity will continue after my death. My apologies and thanks to the nurse or mortician who will deal with the cleanup.

Here's how pooping works when you're alive. Poop follows a winding journey through your body before its final push to freedom. The rectum is the last stop. When it gets there, signals are sent to your brain letting you know, "Hey girl, it's poop time." There is a circular muscle called the external anal sphincter that snuggles around the anus and locks down the fecal prison, preventing poop from leaving our bodies until we're ready. (Except that one time after the spicy tacos.)

The external anal sphincter is a voluntary muscle, which means our brain is actively willing our bums to stay closed. This is also how our brain tells the sphincter to relax when we safely reach the toilet. We appreciate having that control. It's what allows most of us the privilege to walk through the world without pooping randomly like bunnies.

But when we die, our brains no longer send these messages to our muscles. During rigor mortis your muscles seize up tight, but after several days they relax. The good ship decomposition has set sail, and all muscles relax at that point, including the ones that keep poop (and pee, for that matter) inside. So, if you

happen to have feces or urine in the chamber at the moment of your passing, they're now free to go.

I'm not saying everyone is going to postmortem poo. Many older people, or people who have been sick for some time, have eaten almost nothing in the days or weeks leading up to their deaths. When they die, there's just not much waste in there to be released.

As a mortician, I most often encounter a surprise poo when I arrive to pick up a dead body to take it to the funeral home (this is called a "first call"). As a dead body is pulled upright, flipped over—whatever it takes to get the body safely on the stretcher—squeezing occurs, and some feces may escape the body.

But don't be embarrassed, dear corpse! Morticians are used to cleaning up poop, just like new parents are used to changing dirty diapers. It's part of the job.

Besides, forensic pathologists have it way worse in the poop-interaction department. (This is one of the reasons their average yearly salary is roughly $50,000 more than us morticians.) If someone dies mysteriously, their stomach and poop contents can provide important clues. The person performing the autopsy may end up picking through their feces, searching for any anomalies that could explain the death. I'd rather wipe up a small fecal smear while preparing a dead body than pick through a pile of poop like Laura Dern in *Jurassic Park*.

A typical mortician fear is that a dead person will defecate, purge, or leak a bit when the family comes to visit the body. Who wants the final "memory picture" of Grandpa to be a vague eau de poop? Morticians have a host of tricks to prevent this from happening. Entry-level trick: a diaper. This is my preferred method because it's non-invasive. You'll see what I mean in a second. Mid-level trick: an A/V plug. (A/V doesn't stand for audio/visual. It's, um, more graphic than that. I'll let you take that journey of discovery on your own.) The plug is a clear plastic contraption that looks part wine corkscrew, part plastic stopper for a sink or tub drain. Master-level trick: packing the anal canal with cotton and sewing the anus closed. My personal opinion is that this method is a little much, and we should let our corpses poo in peace. I'm happy to share more fecal opinions, so it's a shame no one seems to be asking.

Do conjoined twins always die at the same time?

The problem with the Biddenden Maids is that nobody is sure if they ever existed. It's not that their story isn't well documented. Mary and Eliza Chulkhurst were born (allegedly) in the year 1100 to a family in Biddenden, England. Conjoined twins, they were attached at the hip and the shoulder. They were a feisty pair. Accounts describe them fighting verbally and physically, punching each other during their worst spats. They sound fun—like a medieval reality show! When the twins were thirty-four years old, Mary became ill and died. The family pleaded with Eliza: "We have to at least try to separate you, or you're going to die as well." But Eliza refused to be separated from Mary, her dead sister, saying, "As we came together, we will also go together." Six hours later, Eliza was also dead.

The twins are still celebrated at Easter in their English town, where biscuits imprinted with their image are distributed to lower-income residents. But even with such a well-documented story, the Biddenden Maids may just be that—a story, a legend. If Mary and Eliza were really conjoined at both the hip *and* shoulder, they would be the only recorded case of living twins fused at more than one location.

Though society has an (often inappropriate) fascination

with the secret lives of conjoined twins, they are incredibly rare. We may see them in medical museums and starring in cable television shows, but they just aren't that common—one in every 200,000 births. This type of twin is so rare that scientists still don't fully understand what causes twins to conjoin. The most popular theory is that conjoined twins start as identical twins. Identical twins start as a single fertilized egg that divides in two. If that egg doesn't divide completely, or takes too long to split, then the twins might be conjoined. Another theory believes the opposite: that conjoined twins are two fertilized eggs fusing together.

Though we're not sure how conjoining occurs, we know that when it does happen, the prognosis is . . . grim. Almost 60 percent of conjoined twins will die in the womb before birth. If the twins are born alive, 35 percent won't survive their first day.

If you are one of the rare sets of twins that makes it out of the womb and into the world alive, your chance of long-term survival often depends on where you are joined. For example, if you're joined at the chest or stomach (which most conjoined twins are) and share something like your intestines or liver, you have a much better chance of survival (and are likelier to qualify for separation surgery) than if you are joined at the head.

Conjoined twins born in the twenty-first century are often separated as soon as possible, before the babies are a year old. But even with the best surgeons, in the best hospitals, illness or death for one twin can lead to death for the other twin as well.

Amy and Angela Lakeberg were American conjoined twins born in 1993, sharing a single (malformed) heart and a fused liver. Doctors knew that the girls couldn't survive joined, so the decision was made to sacrifice Amy so that Angela could live. Amy died during the separation, but Angela thrived (for a time). Ten months later, fluid backed up in her heart and Angela died as well. The twins' surgery and hospital care cost over one million dollars.

The island of Malta witnessed a happier ending (although there are no "happy endings" when babies die) in the year 2000. Gracie and Rosie Attard were born sharing a spine, bladder, and much of their circulatory system. Even if conjoined twins have separate organs, like two hearts or two lungs, the organs function in tandem. If one of the twins' organs is much weaker, the other will compensate. Rosie's heart was weak, so Gracie's heart was pumping for both twins. But the strain of pumping so hard threatened to cause Gracie's other major organs to fail. If Gracie's organs failed, both twins would die.

Doctors wanted to separate the twins and sacrifice Rosie, believing that only Gracie was strong enough to survive on her own. But the Attards, Gracie and Rosie's parents, were devout Catholics. They couldn't sign off on "sacrificing" their daughter Rosie, so they chose not to separate the twins at all, and leave things "in God's hands." But a judge, and then an appeals court, ruled against the parents, declaring that the surgery would proceed. Rosie died on the operating table during the twenty-hour separation surgery. Two surgeons both held the scalpel as the aorta was cut, so neither surgeon was solely responsible for Rosie's death. Gracie is now a thriving eighteen-year-old who still keeps in contact with one of the surgeons who performed the operation.

Separating babies can work. It's possible for one baby (and increasingly both babies) to grow and have normal lives. But separation becomes much harder as the twins grow older—

physically, but also mentally. Conjoined twins share an intense bond that not even non-conjoined twins can understand. Adult twins often say they prefer life with their twin. Margaret and Mary Gibb, born in the early twentieth century, had surgeons wanting to separate them since their birth, but they always refused. The demands grew louder through the years, especially after Margaret developed terminal bladder cancer that was spreading to the lungs of both twins. But still the pair refused separation, and the twins died only minutes apart in 1967. They asked to be buried together in a custom casket.

Perhaps the most famous adult conjoined twins were Chang and Eng Bunker. Originally from Siam (now called Thailand), the Bunkers were the origin of the expression "Siamese twins." In their later lives, Chang was unwell, suffering a stroke, bronchitis, and nursing a long-term drinking problem. Eng, it should be noted, never drank. He also claimed not to get drunk or feel any effects from the alcohol that Chang consumed.

One morning, when the twins were sixty-two years old, Eng's son woke the sleeping twins to find that Chang had died. When told, Eng exclaimed, "Then I am going!" and died only two hours later. Scientists believe that Chang died first of a blood clot, and then Eng died as his blood was sent through their connected section to Chang and did not return back to his own body.

It's generally agreed that Chang and Eng could have been separated had they been born in the twentieth century. Today, some hospitals are specifically known for these types of separations. But even state-of-the-art medical technology doesn't guarantee success. In 2003, the Iranian twins Ladan and Laleh Bijani, twenty-nine-year-old lawyers joined at the head, both died during separation surgery. Their surgical team had virtual reality models, CT scans, MRIs, all the latest technology at their disposal. But their fancy systems did didn't detect a hidden vein in the base of the twins' skull. They cut the vein, couldn't stop the bleeding, and the twins died.

Thus the depressing answer to the question "Do conjoined twins always die at the same time?" is "More or less, yes." Sorry, but I don't want to sugar-coat it. Doctors are developing new imaging technology which may help us better understand what's going on deep inside conjoined twins. But the twins are connected in ways (physically and emotionally) that even the latest, most expensive technology will struggle to perceive. Conjoined twins are real people, with real lives and personalities. Well, except maybe the Biddenden Maids. The jury is still out on them.

If I died making a stupid face, would it be stuck like that forever?

We all know the scene: a child running through the house with their eyes crossed, tongue stuck out, and nose pushed up like pig snout. Their long-suffering mother screams after them, "If you keep making that face, it will get stuck like that forever!" Good threat, Mom, but not true. Wacky faces, even the wackiest faces, always pop back into position. (Furthermore, Mom, there is medical evidence all those scrunched, pinched faces are good for circulation.) But what happens if you die making a face? Say, you have a heart attack right in the middle of taunting your mom with an obscene scowl. Will that be your face for eternity?

The answer is mostly no. Intrigued? Read on.

When you die, all the muscles in your body get loose—very loose. (You may recall that this is the time where you might take a small postmortem poo.) This first two-to-three-hour period after death is known as primary relaxation. "Just relax, babe, no worries. You're dead." Even if you happened to be making a silly face when you died, face muscles relax along with everything else during primary relaxation. Your jaw and eyelids will fall open and your joints will get all floppy ("floppy" being the medical term). Say goodbye to your wacky face.

If you or your family is caring for your dead person at the family home or at a nursing facility, our funeral home recommends that the family close the mouth and eyes as soon as possible during primary relaxation. This will set the face in a peaceful position early, before the dreaded rigor mortis begins.

Rigor mortis is more than just the name of a python I used to own. Rigor mortis is the Latin name for the stiffening of the muscles that starts around three hours after death (even sooner in very hot or tropical environments). I've been studying rigor mortis for years and I'm still not sure I totally understand the science of it. The muscles in your body need ATP (adenosine triphosphate) in order to relax. But ATP requires oxygen. No more breathing means no more oxygen, which means no more ATP, which means the muscles seize up and can't relax. This chemical change, collectively called rigor mortis, starts around your eyelids and jaw and spreads through every muscle in the body, even the organs. Rigor mortis makes the muscles incredibly stiff. Once it sets in, that body ain't movin' from whatever position it is in. Funeral directors have to massage and flex the joints and muscles over and over to get them to move, a process called "breaking rigor." This process sounds noisy, full of cracks and pops. But we're not snapping bones; the sounds are coming from the muscles.

Like livor mortis, rigor mortis can be a helpful clue in forensics. A twenty-five-year-old woman in India was found dead, lying on her back. At first glance, investigators might have thought she was a living woman doing yoga or a stretching pose, given that both her legs and one arm were up in the air, seeming to defy gravity. The woman was still stuck in this position when they brought her in for an autopsy. After an investigation, the forensics team developed a theory that the killer may have first murdered the woman and then decided to transport her body to a different location. The killer perhaps placed the woman in this strange position (when she was still in primary relaxation) in order to move her body. During the transport, where she might have been in the trunk of a car or a bag, her body went into rigor mortis. As I explained, once you're in rigor, you're really in rigor. So, when the killer abandoned the woman's body, she was still in the crunched position.

Perhaps we can use rigor mortis to create your post-death silly face? If you asked a friend or relative to set your face in a weird position during primary relaxation, it might get stuck there for the duration of rigor mortis. I'm sure your mom wouldn't appreciate the prank, though. Poor Mom. You're taunting her even in death!

Unfortunately, rigor mortis eventually goes away. Every dead body is different, and the environment plays a big role in the timing, but after about seventy-two hours your muscles will go all floppy again—along with your duck-face lips.

But remember when I said the answer to your question was "Mostly no?" Well here's the rare but fascinating "yes."

There's a controversial phenomenon in forensic science called cadaveric spasm, also known as instantaneous rigor. Instantaneous rigor is exactly what it sounds like. When someone dies, they skip right over the floppy muscle relaxation stage and go straight into rigor mortis. Could this be exactly the loophole we've been looking for to keep your silly face in place during and after your death?

Not so fast. A cadaveric spasm usually affects only one group of muscles, most commonly the arms or hands. This means your arms might get stuck in a funny position after death. Some potential options include zombie arms, "YMCA" arms, or "walk like an Egyptian" arms. But I don't know that zany "after death arms" have quite the same impact as a zany "after death face," like tongue-out-googly-eyes or pig-snout-crossed-eyes.

Also, cadaveric spasms usually follow a stressful death. We're talking seizure, drowning, asphyxiation, electrocution, gunshot wound to the head. They've been observed in soldiers who were shot in battle, or people who died following a brief period of intense struggle. It doesn't sound like a chill situation, and frankly I don't want that kind of bad death for you, my young friend.

I don't see any way for your silly face to be stuck like that forever. I tried to make it work, but the science just isn't there. Besides, you should stop tormenting your poor mother.

Can we give Grandma a Viking funeral?

Did Grandma want a Viking funeral? If so, your grandmother sounds rad and I wish I had known her.

I'm afraid I have some terrible news. Not only is Grandma dead, but "Viking funerals," at least the Hollywood version of them, aren't real. You're picturing Grandma, the fallen warrior, her shrouded body laid solemnly upon her wooden boat. Your aunts push the noble craft into the sea. Your mom draws her bow, a flaming arrow arcs through the sky, hits Grandma, and sets her alight. She burns as bright in death as she did in life.

Alas, fake fake fakeity fake.

How can it be fake? It's called a Viking funeral because, duh, that's what the Vikings did. Well, no. The Vikings, everyone's favorite medieval Scandinavian raiders 'n' traders, had diverse and interesting death rituals, but a flaming cremation boat wasn't one of them. Here are a few rituals that did happen. Vikings performed cremations—on land. Sometimes the cremation pyre would be built inside stones that were outlined and stacked into the shape of a boat (which might be where this idea came from). If the dead person was especially important, their whole boat would be hauled up on land and used as a coffin, known as ship or boat burial. But no flaming-arrow cremation cruise.

As a warning, any time you try to tactfully bring up the historical inaccuracy of someone's blazing boat corpse idea, you'll

be hearing from the "Ahmad ibn Fadlan guy." The Ahmad ibn Fadlan guy is the person on the internet who insists that Hollywood versions of flaming boat cremations are real. AiF-guy spends a lot of time making this argument, and he supports his case with the writings of a man named Ahmad ibn Fadlan, an Arab traveler and writer from the tenth century. Ahmad ibn Fadlan is known for documenting what he called the Rus'—the northern Germanic Viking traders. Ibn Fadlan is a problematic historical source, in part because he was a biased observer.

For example, he thought the Vikings were "perfect physical specimens," but was openly hostile about their hygiene. His chronicles mention an elaborate cremation ritual the Rus' performed for one of their chieftains.

According to Ibn Fadlan, the Rus' stored their chieftain in a temporary grave for ten days. Because the chieftain was so important, his people pulled his entire longship ashore and hauled it onto a wooden platform. An older woman, who was in charge of the ritual and was known as the Angel of Death (hold on, Ibn Fadlan: I want to hear more about this Angel of Death woman), made a bed for the chieftain on the boat. The chieftain was taken out of his grave, re-dressed, and placed on the bed with all his weapons around him. His relatives arrived with flaming torches and set the boat alight, and the whole thing plus wooden platform burned along with him. Important: this all happens on land.

Who knows how this whole rumor got started. The Vikings had elaborate cremations! They had boats! They just didn't have cremation boats.

I know what you're thinking. "Okay, fine, so my death plan is a little historically inaccurate. I wasn't that into Norse history anyway. Let's go ahead with the flaming boat thing!" Not so fast, my postmortem pyro. The reason that no culture has adopted the flaming boat funeral custom is because it doesn't work.

I've seen an open-air funeral pyre. The first fifteen minutes after the fire is lit are jaw-dropping. Smoke curls around the corpse and red-hot flames shoot up from the body. You can see why Hollywood would say, "We love this fiery pyre scene, but—stay with me—what if it were on a boat?" Here's the thing: after those first fifteen minutes of glorious flame, you still need several hours and a lot of wood to fully cremate that body. Your average canoe is between sixteen and seventeen feet. It could carry enough wood to start the pyre off, but I have it on good authority (the cremation pyre people told me) that a full cremation requires over 40 cubic feet of wood. The fire has to reach 1200 degrees and stay there for two to three hours. You have to keep adding wood close to the body throughout the cremation. Even stacked high with logs, a sixteen-foot Viking boat holds nowhere close to the amount of wood needed. The fire would most likely burn a hole in the boat before it got hot enough to burn the corpse, so the whole setup is still very inefficient. When the death boat burns out too quickly, what does that leave us with? A half-charred body bobbing around your local municipal waterway. The historical romance would be ruined if Grandma's body washed up on shore during someone's family picnic.

I know this is bad news, and I hate to be the mortician bringing all the bad news. So here are some things you can do instead.

One: Have Grandma cremated in an ordinary cremation machine, called a retort. You can watch Grandma's body being loaded into the machine and blast Norse battle chants while you press the button to start the flames. This is called a witness cremation. Then, you can take her cremated remains and put them on a tiny Viking boat and set that on fire, sending it out into a body of water. As the wee boat burns, the ashes will scatter into the water. (Note: I'm not advocating anyone set things on fire in public waterways. I'm just saying, hypothetically, it might be cool.)

Two: Make sure your grandma's fingernails and toenails are nicely trimmed before she's cremated. According to Norse lore, a bunch of dark stuff called Ragnarök is going to go down, ending in a huge battle where the gods die and the world is destroyed. During the battle, a vengeful army will arrive on a giant ship called the Naglfar, or nail ship. That's right, an entire battleship made out of the fingernails and toenails of dead people. So if you don't want Grandma's nails to contribute to the fall of the universe, get out the clippers and snippy-snip. If you follow these steps, sure, it's still not a "Viking funeral," but at least you get the flaming boat and a heroic manicure.

Why don't animals dig up all the graves?

It depends what type of grave you're talking about. When you bury a family pet—like your cat, your dog, or your fish (if it escaped the flush to the great beyond)—it's possible that another wild animal, like a coyote, might dig that grave up. The coyote isn't participating in ritual grave desecration, it's just looking for a free meal. Listen, it's not the coyote's fault that your family dug that backyard hole for Fido only a foot deep. (Psst—not deep enough.)

As an animal begins to decompose beneath the soil, it produces some very pungent-smelling compounds called cadaverine and putrescine. Compounds named for "cadaver" and "putrid"—adorable, right? To a scavenger animal, those decomposition compounds smell a lot like dinner. If they sense their meal is an easy dig away, they might go for it.

There's a simple fix here: shovel down a little deeper (I'll reveal just how deep in a second) for Fido's eternal resting spot.

But what about human cemeteries? Cemeteries exist in almost every town, but one rarely sees scavenger animals roaming around, digging up fresh bodies for food.

That's not to say it's impossible. In remote parts of Russia and Siberia, armed guards have had to stand watch at cemeteries after black and brown bears broke in to dig up human remains. In one memorable story, two village women thought they were seeing a man in a big fur coat bending down to tend a loved one's grave. Wrong: it was a bear eating a dead body it had dug out of the ground. Sorry, ladies.

Another recent story from Bradenton, Florida, involves neighbors noticing dog or coyote tracks around half a dozen graves at a local cemetery. Fresh holes had been dug, releasing a gnarly smell. Body bags poked out of the ground.

I mention these two horror stories to make an important point: they are the exceptions that prove the rule. For the most part, animals won't dig up human graves. There are several reasons why. First, the correct amount of soil laid on top of the body creates a scent barrier. Second, the soil not only covers up the powerful smell, but it actively works to decompose the body, leaving behind a stenchless skeleton. The soil is magic.

The real question is, how deep is deep enough for a grave? Just to be safe, shouldn't we bury all our humans six feet under in the heaviest caskets we can make, and further fortify them in underground concrete bunkers? No. Because the soil's magic benefits are most magical (scientific terms here) near the surface. This is where you find the highest numbers of fungi, insects, and bacteria that efficiently decompose a body from human to skeleton. If you bury a body too far down, the soil grows sterile. Topsoil has more oxygen, which means your body can become a tree . . . or a bush . . . or a shrub at the least. To become "one with the earth," you want to be as close to the surface as you can.

So what's the compromise here? There are those who argue that a body needs to be buried a full six feet down, but there are also those who argue that just one foot of soil is needed to create a smell barrier. I think three and a half feet is a good compromise. "Three and a half feet, you won't become a treat!"

as the old saying goes. (This is not an old saying, FYI.) That depth puts at least two feet of smell barrier above the corpse, while maintaining the benefits of that awesome decomposing soil near the surface. Three and a half feet is the standard at natural burial grounds across the United States, and there have been zero reports of animals digging up graves.

Full disclosure: even if you're buried under two to three feet of dirt, it's still possible that animals may get a whiff of you. Every once in a while, animal tracks (like coyote) are spotted around a gravesite, as if to say, "Well well, what have we here?" But they don't dig up the grave because it's too much damn work. Think of it like this: why do I get Taco Bell from the drive-thru instead of cooking myself a spinach kale casserole for dinner with organic ingredients from the farmer's market? If a scavenger animal can get food elsewhere, it's not worth it for him to dig two feet down into the soil and try to haul up your giant human butt. Scavengers have other things to worry about, like protecting their territory and themselves. They don't have the time and energy to dig that massive hole just to chew on your femur. Besides, animals like coyotes and bears aren't physically well suited to digging that deep.

So, why were those bears in Siberia messing around in the graveyard? I suspect that the graves weren't deep enough. The ground is often frozen that far north. If it's easier for a bear (who, remember, doesn't have great digging paws) to dig up Grandpa's corpse than it is for the bear to hunt, the graves aren't deep enough. Second, and much more important, the bears were starving. Their mushrooms and berries (and the occasional frog, apparently) that make up their normal diet were in short supply. The bears began by raiding the cemetery for food that families left as offerings at the graveside. They were eating everything from cookies to candles, whatever they could find to stay alive. Only after raiding that easy-to-access food did the bears turn to digging up the bodies.

And what about the cemetery in Florida? In this older, abandoned cemetery, why were there fresh graves, terrible smells, and body bags? It turns out that the graves had been dug by a local funeral home to bury homeless people. And because the "abandoned" cemetery didn't have government oversight, the funeral home is alleged to have buried the bodies in extremely shallow graves. Since then, the director has put cement slabs over the graves. Good thing there are no bears in Bradenton, Florida!*

I'll end with a lesson about badgers digging up medieval bones. Back in the Middle Ages, people used to be buried right outside (and even inside) churches—lots and lots of people. The human remains were supposed to have been moved away from one particular thirteenth-century English church back in the 1970s. But it turns out they weren't all moved. We discovered this because badgers invaded and started digging dens and networks of tunnels down through the ancient bones, sending pelvises and femurs flying to the surface. Someone should stop those badgers! Whoops, they can't. In England it's illegal to kill these furry creatures, or even move their dens. Thanks to the Protection of Badgers Act (yes,

* Turns out there are, but they're *very* rare.

that's real), we're looking at six months in prison and huge fines even for attempted badger assault. Workers at the church have to pick up the bones, say a prayer, and bury them back in the ground. The lesson here is that even if you make it almost a thousand years in the grave, you never know when you'll be uprooted by a lawless badger.

What would happen if you swallowed a bag of popcorn before you died and were cremated?

I have a sneaking suspicion you asked this question because of a meme that's been everywhere over the last few years. It's a picture of a bag of movie popcorn, with the words, "Right before I die, I am going to swallow a bag of popcorn kernels. It will make the cremation epic."

I get it. You want to be unique, to stand out, even when you're dead. You're a whimsical prankster, Tim! It would just be "soooooo Tim" to swallow popcorn kernels before you die. That way, when you're put in the cremation machine, the popcorn kernels will pop like fireworks and come pouring out of your corpse and the crematory operator will leap away in shock before admitting, "That was soooooo Tim! You sure got me, Tim."

Listen, not gonna work, Tim. For many reasons. First of all, you think you're going to be on your deathbed, weak, organs failing, not having eaten solid food for weeks, and all of a sudden you're going to be in the mood to smuggle a bag of popcorn kernels into the nursing home and then swallow what amounts to a bowlful of tiny marbles? "Sorry, honey, as much as I want to whisper my final 'I love you' as I take my last breath, first I have to gobble down this bag of PopSecret." Probably not.

Even if you did manage to swallow a whole bag of popcorn, are you clear on how a cremation machine works? This meme  has become popular because most people don't know what a crematory looks like, sounds like, or how the process goes. In order for the popcorn prank to work, you have to believe that Tim's body would burst open mid-cremation, releasing all these popcorn kernels. Also, that a single bag of microwave popcorn is going to create waves and waves of popcorn, like when pranksters put soap into decorative fountains at their high school and suds overflow into the courtyard. (By my calculations, you'd need to swallow at least a gallon and a half of unpopped kernels to create an impressive wave once popped.) The other part of the joke is that the deafening popping noise will shock the crematory operator into thinking the crematory is under attack.

Here are two reasons that will never happen. (There are countless reasons it will never happen, but let's focus on these two.)

One: Cremations take place in fourteen-ton machines with huge burners and combustion chambers, and a thick metal door that seals the dead body inside the brick chamber. The cremation machine is loud. Real loud. Even if you had forty-seven bags of popcorn in there with you, you'd never be able to hear them popping from the outside.

Two: More importantly, even if you could hear the popcorn popping, it doesn't matter 'cause it's not gonna pop! What's the main complaint everyone has about popcorn? All the unpopped and burnt kernels at the bottom. Conditions have to be ideal to pop a delicious bowl of popcorn. The kernels have to be just the right amount of dry, which they definitely would not be if they had been hanging out digesting in your stomach, a moist compressed environment.

Researchers (engineers using thermodynamic analysis . . . I'm serious) discovered that the ideal temperature for popping popcorn is 356 degrees Fahrenheit. If you're popping the kernels in oil on the stove, the oil should be just over 400 degrees. If the temperature creeps up much higher, the popcorn will burn before it has the chance to pop. An average temperature for a cremation machine is 1,700 degrees Fahrenheit. That's more than three times the temperature popcorn should be popped at. Plus, a column of flame shoots down from the ceiling and hits the chest and stomach. Those kernels would just blacken and disappear without a trace—like the other soft tissue in your body.

I don't feel that bad about ruining your prank here, Tim, because why were you trying to trick the crematory operator in the first place? As someone who worked as a crematory operator in my twenties, I can tell you it's a hard job. It's dirty, hot, and you're hanging out with dead bodies and weeping families all day. The crematory operator doesn't need your tomfoolery, Tim!

But if you're dead set on creating explosions that a crematory operator would hear and would definitely be freaked out by, don't leave unpopped popcorn in your body. Instead, try leaving a pacemaker in your body. (Note: I one-thousand-percent do not recommend doing this. I'm making a joke. See, I can make jokes too, Tim.)

A pacemaker helps living people control their heartbeat,

speeding up the heart if needed, slowing the heart down if needed. It's a cute lil' thing, the size of a small cookie, that is basically a battery, generator, and some wires implanted (through surgery) into the body. It can save your life if your heart is misfiring. But if a pacemaker is not removed from a dead body before the cremation, it can turn into a tiny bomb.

Before I put a dead body into a cremation machine, not only do I check the paperwork to see if the person had a pacemaker, I also poke around a bit in the area above the heart. If there is a pacemaker, it has to be cut out of the body. Don't worry, the person is dead, they don't mind. Pacemakers aren't exactly rare, either. Over 700,000 people a year get them. So it's not surprising that some pacemakers slip through and slide into the machine with the body.

If that happens, the very high heat can cause a flammable chemical reaction that makes the pacemaker burst. All the energy in the battery? The energy that is supposed to power the pacemaker for years? Bam! It gets released in one second. There's an explosion, which can terrify or maim the crematory operator, especially if they happen to be peeking into the machine to check on the cremation at the time. The explosion can also break the door of the machine or damage the bricks inside.

I hope you never need a pacemaker, Tim. But I also hope your postmortem pranks are a little tamer. Maybe schedule a tweet to go out two weeks after you die? One that says, "Every step you take I'll be watching you." That'll get 'em.

If someone is trying to sell a house, do they have to tell the buyer someone died there?

As I write, there are some brand-new luxury condos being built in my neighborhood in Los Angeles. They're overpriced and not very attractive (think, giant white Tupperware), but we can be pretty sure no one has died in them. Yet.

Pro Tip: if you have your heart set on living somewhere that absolutely, positively no one has ever died in, buy a new house. Preferably one that you've watched being built. Because the truth is, if you live in a charming prewar bungalow or a grand Victorian mansion, it's possible you're watching TV and eating popcorn where somebody breathed their last breath. And nobody has to tell you about it.

The laws differ from place to place on what someone selling a house is legally required to tell the buyer. Generally speaking, if someone died a "peaceful death" in a home (meaning it wasn't part of an ax murderer's chopping spree), the seller doesn't have to tell the buyer. The same goes for accidental deaths (say, falling off a ladder) and suicides. And no place in the United States requires sellers to disclose deaths related to HIV or AIDS. In some cases, the seller will be advised not to reveal that a death has occurred, as it could cause unnecessary stigmatization of the property. No seller wants the buyer's mind reeling off into visions of gory crime scenes, torrents of blood like the elevator in *The Shining*, or you know, ghosts.

Death has happened in many homes, more homes than you probably realize. *Perhaps in the very house in which you're reading this book.* Remember, people mostly used to die in their own homes, not at hospitals or nursing homes, so if your house has been around for one hundred years or more, it's highly likely to have seen death within its walls.

If someone died peacefully in their home, they were probably attended by loved ones or hospice workers. After the death, the corpse was removed from the home way before any heavy-duty putrefaction set in. These are not the types of deaths that ghost stories are made of.

Even if, for some reason, there *was* heavy-duty putrefaction going on, a skilled cleanup crew can get a place so spick and span that you'd never know that there was once a corpse decomposing in the room that's now your man cave.

For example, a friend of mine, I'll call her Jessica, lived in a fifth-floor apartment in Los Angeles. One spring she noticed an odd smell pervading her apartment. At first, she thought she just needed to do a better job of cleaning her cat's litter box.

It wasn't long before it became clear that the smell was coming from the apartment directly underneath hers. A man had died alone at home, and nobody had found his body for over two weeks. The "cat litter" smell was decay, wafting up through the floorboards of the old apartment building. Authorities were called, and the corpse was removed.

Jessica, unable to help herself, climbed down the fire escape to peek in the open windows of the dead man's apartment. She saw what remained of her neighbor after the coroner took the body. A thick, black stain spread across the floor, and rogue maggots wriggled through the liquid.

No, you obviously wouldn't want to rent the apartment in that condition. But, fast-forward a couple of months, and the apartment had been overhauled—everything was sparkling clean—and rented again. Jessica met the new people who moved in and asked them how they liked their new apartment. They were very happy, no complaints about smells or such. Jessica decided not to say anything about her former neighbor.

Did the new tenants know a death had occurred in their apartment? Legally, a California landlord has to tell you if there has been a death in the apartment within the last three years. California is one of the only states with a law this specific. If, later on, the tenant comes to feel harmed by the death in their home, they may be able to sue. So disclosing the death to them in advance, before they rent, is really the only way for the landlord to protect themselves from a future lawsuit. But it's possible that Jessica's landlord didn't know the law (or ignored it) and never said anything.

It's worth noting that in some U.S. states, Georgia for example, a landlord only has to tell you about a recent death *if you ask*. But if you do ask, they are required to answer truthfully. Sort of like how a vampire can only come into your house if you invite them in. The takeaway from Jessica's story is, if you are worried about recent deaths in your potential new home, you should ask.

Asking should work in most places, but not all places. (Oregon, I'm looking at you.) In Oregon, it doesn't matter when or how someone died; nobody has to tell you anything. Brutal, violent deaths included. Murder, suicide, peaceful death—it's all the same in the Beaver State.

In realtor-speak, what matters is something called "material facts." Material facts are things that can affect a buyer's desire to purchase a property. Most often this is stuff like cracks in the foundation or invisible structural problems. Depending on what state you're in, a violent death, like a murder, might fall into the category of material fact, meaning it has to be disclosed. But peaceful or accidental deaths are not usually considered material facts.

Being the site of a grisly murder can turn a house into a "stigmatized property"—that is, a house with a "reputation." The same goes for reports of violent crimes, or even hauntings. The seller probably doesn't want to tell you about the triple homicide there in 2008, but if they don't tell you and you learn about it from the neighbors (the house has a "reputation," you see), you might have grounds to terminate the contract or file a lawsuit down the line. Again, it probably depends on what state you're in.

Really, the best thing I can tell you is to get comfortable with the fact that you may someday live in a house or apartment where a person died. You'll be okay. My mother is a realtor, and just sold a house in which the ninety-year-old previous owner died. Mom told the potential buyers (because she knew the neighbors would tell them if she didn't), and they went home to think about it. They came back and wanted the house anyway, because the woman must have loved the house so much she wanted to die there.

I hope to die peacefully at home, and I don't plan on staying around to haunt it, either. But if you're still terrified that someone has died in your next potential domicile, get comfortable having those types of conversation with your realtor or landlord.

Unless you're in Oregon.

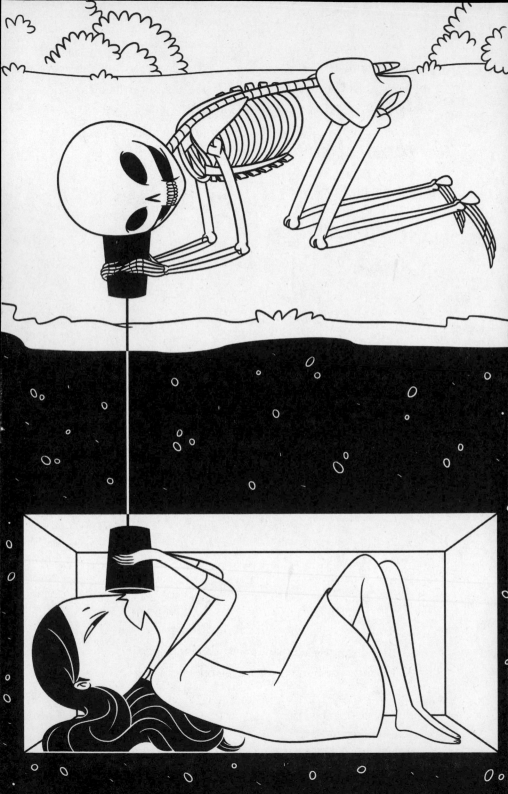

What if they make a mistake and bury me when I'm just in a coma?

Okay, so to be clear, you *don't* want to be buried alive, is that correct? Got it.

Lucky for you, you don't live in Ye Olden Times! During Ye Olden Times (before the twentieth century), doctors had a less-than-flawless track record when it came to declaring people dead. The tests they used to determine if someone was honest-to-God-really-dead were not just low-tech, they were horrifying.

For your enjoyment, here's a fun sample of the death tests:

* Shoving needles under the toenails, or into the heart or stomach.
* Slicing the feet with knives or burning them with red-hot pokers.
* Smoke enemas for drowning victims—someone would literally "blow smoke up your ass" to see if it would warm you up and make you breathe.
* Burning the hand or chopping off a finger.

And, my personal favorite:

✶ Writing "I am really dead" in invisible ink (made from
   acetate of lead) on a piece of paper, then putting the
   paper over the corpse-in-question's face. According to
   the inventor of this method, if the body was putrefying, sulfur dioxide would be emitted, thus revealing
   the message. Unfortunately, sulfur dioxide can also
   be emitted by living people, like those with decaying
   teeth. So, it's possible there were a few false positives.

If you woke up, breathed, or visibly responded to these "tests"—
hallelujah!—you weren't dead. But you might be maimed. And
that needle stuck in your heart could actually kill you.

But what about the poor souls who weren't put through the
battery of stabs, slices, and enemas, but were just assumed to be
100 percent dead and sent to the grave?

Take the tale of Matthew Wall, a man living (yes, *living*)
in Braughing, England, in the sixteenth century. Matthew was
thought to be dead, but was lucky enough to have his pall-
bearers slip on wet leaves and drop the coffin on the way to
his burial. As the story goes, when the coffin was dropped,
Matthew awakened and knocked on the lid to be released. To
this day, every October 2nd is celebrated as Old Man's Day
to commemorate Matthew's revival. He lived, by the way, *for
twenty-four more years.*

With stories like that, it's no wonder that certain cultures
had extreme taphophobia, or the fear of being buried alive.
Matthew Wall was lucky that his "body" never reached his
grave, but Angelo Hays was not.

In 1937—true, 1937 is not quite Ye Olden Times, but at least
it's way before you were born—Angelo Hays of France was in a
motorcycle accident. When doctors couldn't find his pulse, he

was pronounced dead. He was buried quickly and his own parents were not allowed to see his disfigured body. Angelo would have remained buried if it wasn't for the life insurance company's suspicions of foul play.

Two days after Angelo was buried, he was exhumed for an investigation. Upon inspecting the "corpse," examiners found that it was still warm, and that Angelo was alive.

The theory is that Angelo had been in a very deep coma which slowed his breathing way, way down. It was that slow breathing that allowed him to stay alive while buried.* Angelo recovered, lived a full life, and even invented a "security coffin" with a radio transmitter and a toilet.

Luckily, if you fall into a coma today, in the twenty-first century, there are many, many ways to make sure that you are good and dead before you're moved on to burial. But while the tests may show that you are technically alive, your new status may be small comfort to you and your kin.

Media and TV shows often throw around terms like "coma" and "brain-dead" interchangeably. "Chloe was my true love, and now she will never wake from her coma. I must decide whether to pull the plug." This Hollywood version of medicine can make it seem like those conditions are the same, just one step away from death. Not true!

Of the two, the one you really don't want to be is brain-dead. (I mean, neither is great, let's be honest.) But once you're brain-dead, there is no coming back. Not only have you lost all the upper brain functions that create your memories and behaviors

---

* If you're buried alive and breathing normally, you're likely to die from suffocation. A person can live on the air in a coffin for a little over five hours, tops. If you start hyperventilating, panicked that you've been buried alive, the oxygen will likely run out sooner.

and allow you to think and talk, but you have also lost all the involuntary stuff your lower brain does to keep you alive, like controlling your heart, respiration, nervous system, temperature, and reflexes. There are gobs of biological actions controlled by your brain so that you don't have to constantly remind yourself, "Stay alive, stay alive . . ." If you are brain-dead, these functions are being performed by hospital equipment like ventilators and catheters.

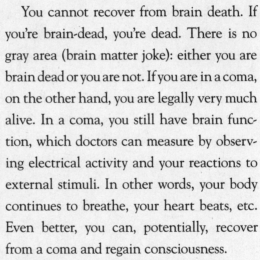

You cannot recover from brain death. If you're brain-dead, you're dead. There is no gray area (brain matter joke): either you are brain dead or you are not. If you are in a coma, on the other hand, you are legally very much alive. In a coma, you still have brain function, which doctors can measure by observing electrical activity and your reactions to external stimuli. In other words, your body continues to breathe, your heart beats, etc. Even better, you can, potentially, recover from a coma and regain consciousness.

Okay, but what if I fall into a deep, deep coma? Will someone eventually pull the plug and send me off to the mortuary? Will I be trapped in both a casket and in the *prison of my mind*?

No. We now have a whole battery of scientific tests to confirm that someone is not just in a coma, but really, truly brain-dead.

These tests include but are not limited to:

✳ Seeing if your pupils are reactive. When a bright light is shined into them, do they contract? Brain-dead people's eyes don't do anything.

* Dragging a cotton swab over your eyeball. If you blink, you're alive!

* Testing your gag reflex. Your breathing tube might be moved in and out of your throat, to see if you gag. Dead people don't gag.

* Injecting ice water into your ear canal. If doctors do this to you and your eyes don't flick quickly from side to side, it's not looking good.

* Checking for spontaneous respiration. If you are removed from a ventilator, CO_2 builds up in your system, essentially suffocating you. When blood CO_2 levels reach 55 mm Hg, a living brain will usually tell the body to spontaneously breathe. If that doesn't happen, your brain stem is dead.

* An EEG, or electroencephalogram, which is an all-or-nothing test. Either there is electrical activity in your brain or there isn't. Dead brains have zero electrical activity.

* A CBF, or cerebral blood flow study. A radioactive isotope is injected into your bloodstream. After a period of time, a radioactive counter is held over your head to see if blood is flowing to your brain. If there is blood flow to the brain, the brain cannot be called dead.

* Administering atropine IV. A living patient's heart rate will accelerate, but a brain-dead patient's heartbeat will not change.

A person has to fail *a lot* of tests to be declared brain-dead. And more than one doctor has to confirm brain death. Only after countless tests and an in-depth physical exam will you go from "coma patient" to "brain-dead" patient. Nowadays, it's not just some dude with a needle poised over your heart and "I am really dead" scrawled on a scrap of paper.

It is highly unlikely that your living brain will slip through the cracks and that you'll be sent away from the hospital in a coma. Even if you were, there is no funeral director or medical examiner I know who can't tell the difference between a living person and a corpse. Having seen thousands of dead bodies in my career, let me tell you—dead people are very dead in a very predictable way. Not that my words sound all that comforting. Or scientific. But I feel confident saying that this is not going to happen to you. On your list of "Freaky Ways to Die" you can move "buried alive—coma" down to just below "terrible gopher accident."

What would happen if you died on a plane?

The flight attendant would open the plane's emergency exit door and toss your body out, attached to a parachute. Before you head out the door, they'd place a little card in your pocket that lists your name and address and says, "Don't worry, I'm already dead."

(I'm being informed by fact-checkers that this is not official airline policy.)

If you die on a plane, it's usually not because the plane has crashed. Plane crashes are very rare; your chances of being in a plane crash are one in 11 million. I tell you this statistic because personally I am freaked the heck out by plane crashes. But it's just not gonna happen. You're safe up there.

But with 8 million people flying every day, it is almost inevitable that someone will die from heart problems, lung problems, or other ailments related to old age. Dying somewhere over the Atlantic Ocean after a complimentary ginger ale is always a possibility. A few years ago, I was flying from Los Angeles to London. After our chicken tikka masala dinner, the guy next to me keeled over into the aisle, puking up his tikka masala and lying completely motionless. "Oh crap, this is not a drill!" I thought. As a mortician, I wasn't good for much beyond being comfortable sitting next to a dead person the rest of the way to London. Fortunately, there was an actual doctor on board. She got the gentleman back up and running, and he even got to sit in first class for the rest of the flight. (I was back in coach with the lingering tikka masala puke smell.)

The flight crew will respond in different ways depending on whether there's been a medical emergency or a death on the flight. If the person is clinging to life and can still be saved, the flight crew will try to divert the plane and land at the airport nearest to medical personnel and a hospital. But if the person dies? Well, they're dead now, and they're still going to be dead when we land in Bora Bora. What's the rush?

If you happen to be the one sitting next to the person, you will find yourself living through the undeniably surreal experience of having a dead seatmate. "Excuse me," you'll say to the flight attendant, "I am sorry to bother you, but I didn't sign up to sit next to a corpse for the remaining five hours of the flight." Especially if you're trapped in the window seat, while the dead guy has the aisle seat. But no worries, the flight crew will whisk the body away immediately and store it out of sight, right?

Uh, no. They will 100 percent leave it in the seat next to you.

In the more glamorous days of air travel, airlines would always leave several seats open, which would allow for a corpse to at least have its own row. But nowadays, any frequent flyer knows that airlines pack their flights completely full. If that's the case, the flight attendant might drape one of those scratchy blue airline blankets over the dead person, buckle them in, and call it a day.

"Surely there must be some secret place on the airplane to store a dead body," you say. Have you been on a plane? We're packed like sardines in there. The airplane bathroom is not an option. The person will slump down to the floor, making it impossible to open the door upon landing. If the flight is longer than three hours, rigor mortis might set in, making removal even more challenging. Plus, sticking Grandma into the plane lavatory is not particularly respectful. Remaining available options are: body in an empty row (if one is available), body in the seat next to you (if zero other seats are available), or body in the back galley (where the beverage carts come from). Best-case scenario, the flight attendants might lay the body in the galley, cover it up, and close the curtain.

Once upon a time (like, 2004), Singapore Airlines actually installed the secret corpse cupboards we assume all airlines have. Aware that people died while flying, the airline was attempting to "take the trauma out of such tragedies." The cupboards, complete with straps so that the body didn't go soaring on a bumpy landing, were built into their Airbus A340–500. This particular aircraft was used for the longest flight in the world at the time— seventeen hours, from Singapore to Los  Angeles—with very few places to land along the way. Sadly, these Airbuses have since been discontinued, along with their revolutionary corpse cupboards.

You probably don't love the idea of a dead body on your flight. I am extremely comfortable with dead bodies, but even I could do without sitting for several hours next to a stranger's corpse. But would it make you feel better if I told you there are dead bodies on your flight often, you just don't realize it? I'm talking about bodies in the cargo hold of the plane, down with your luggage. Dead people are zipping from one place to another all the time. Say the dead person lived in California, but wanted to be buried in Michigan. Or the person died on vacation in Mexico, but has to be brought back to New York. We handle bodies like this at my funeral home all the time. We pack them very safely in heavy-duty flying cases, drop them off at the airport, and send them soaring away home. Any flight you take, there might be an extra passenger tucked below.

As a final note: according to the flight crew, no one ever really dies on a plane. If a person were to die mid-flight, that would mean a bunch of hassle and paperwork for the crew. The whole flight could even be quarantined upon landing for fear of disease. Then there's the possibility that the police will consider the plane a potential crime scene and take it out of commission while they investigate. It's hard enough to make airline connections without an episode of *Law & Order* happening at seat 32B. Rather than admit death in the sky, the protocol is to ask that medical personnel declare the person dead once on the ground. Most flight attendants aren't doctors and can argue that they aren't qualified to declare a passenger legally dead. Sure, the passenger hasn't breathed for three hours and they're in rigor mortis, but that doesn't prove anything!

So now you know what to expect if someone dies on your plane. Sitting next to a corpse all the way to Tokyo isn't ideal, but I would prefer a corpse to a crying baby. No offense to babies, I just spend more time around corpses.

Do bodies in the cemetery make the water we drink taste bad?

Wait a second. What do you have against a tall, delicious glass of corpse water?

All right, fine, no one wants corpses near their water. The thought is gross, no matter how death-accepting you are. Every once in a while, we hear a horrible story about dead bodies contaminating the water supply somewhere in the world. Cholera is a perfect example—a disease that you really don't want to get. Cholera is spread through a poop cycle: the bacteria that causes cholera gets into your intestines and gives you horrendous watery diarrhea for days on end. Left untreated, it can kill you. If that horrendous watery diarrhea gets into the water supply, it will create unsafe drinking water, which then causes more cholera. Around 4 million people around the world are infected every year, often poorer people living in places that lack clean water.

Where do corpses come in? Well, in places like West Africa, cholera outbreaks have been caused by dead bodies without

people being aware of it. When a beloved family member dies of cholera, their relatives will wash and prepare the body. The feces from the corpse (contaminated with cholera) get into the water, or are transferred on the hands of the body washers who go on to prepare the funeral feasts. The water and food served at the feast are contaminated with the bacteria and, before you know it, cholera outbreak.

This sounds terrifying, but I want to be clear: only very specific infectious diseases (like cholera and ebola) can make a dead body dangerous in any way. These are diseases that are extremely rare right now in places like the United States and Europe. You're more likely to die from your pajamas catching on fire than from ebola. And we are incredibly lucky to have expensive sanitation and waste systems that neutralize cholera. If you want to wash and care for a body that has died of cancer, a heart attack, or a motorcycle accident, and then turn around and make a feast of drinks and food, everyone, washers and eaters, will be safe. (I still recommend you wash your hands before doing any food prep, whether or not your day includes corpse-handling.)

What about when a whole dead body is in the water? This is a more extreme example, of course. For the ick factor alone, no one wants a human corpse or fresh skunk carcass floating in their water supply. But what about bodies buried in a cemetery? Bodies are decomposing underground, and underground is where rural communities source their water. Decomposition seems pretty disgusting. It can't be good to have rotting bodies near the water we drink, right?

Scientists have done studies on this exact issue, and have answers for you.

Decomposition may look (and smell) disgusting, but the bacteria involved in decomposing a dead body aren't dangerous. Not all bacteria are bad. These are friendly bacteria that don't cause disease in living people; they just chomp away at dead ones.

To learn what happens to bodies after burial, scientists study decomposition products ("decomposition products" make me think of branded T-shirts and iPhone cases) in the water and soil around a grave. If buried only a few feet from the surface, a body that hasn't been chemically preserved will decompose very quickly. The rich soil acts as a "purifying element that shortens the decomposition period." Not only that: this soil close to the surface will keep contamination from going deep into the soil where the water is. As long as the body doesn't have one of those highly infectious diseases I mentioned, the water should be fine.

In fact, the things we do to dead bodies to prevent them from decomposing can do more damage than just letting the bodies decompose naturally. Often buried bodies are placed in a thick hardwood or metal coffin, chemically preserved, and buried very deep, six feet or more down in the soil. The idea is that it's safe down there, for both the body and everyone else. But the metals, formaldehyde, and medical waste may do more harm to the groundwater than the body they're trying to protect.

For example, did you know that Civil War soldiers are still attacking . . . the water supply? It's strange, but true. Over 600,000 soldiers died during the Civil War, and their grief-stricken families wanted their bodies brought home for burial. But stacking the rotting corpses in train cars and shipping them home wasn't an option (the exasperated train conductors weren't having it). And most families couldn't afford the expensive iron coffins that the train companies would allow. So enterprising men, called embalmers, started following the armies around, setting up tents, and chemically preserving the soldiers killed in battle so that they wouldn't decompose on the journey back home. The embalmers, who were still experimenting with their craft, used everything from sawdust to arsenic. The prob-

lem with arsenic is that it's toxic to living humans. Extremely, wildly toxic—causing various cancers, heart disease, developmental problems in babies . . . the list is long. And 150 years after the Civil War ended, deadly arsenic is still seeping from the ground in Civil War–era cemeteries.

As the soldiers slowly decompose below ground, their bodies mix with the soil, releasing arsenic. As rain and floodwater move through the soil, concentrated clumps of arsenic are washed into the local water supply. Any amount of arsenic in your water is too much arsenic, frankly—but in trace amounts it's safe to drink. Still, a study at a Civil War cemetery in Iowa City found that nearby water contained arsenic at three times the safe limit.

It's not the soldiers' fault. Their decomposing bodies wouldn't cause cancer if we hadn't stuffed them full of arsenic. Fortunately, embalmers stopped using arsenic over a hundred years ago—though formaldehyde (arsenic's replacement) isn't without its own toxic problems.

Again, unless you are washing a body that died of ebola or cholera (probably not), or live next door to a Civil War–era cemetery (slightly more likely, but still, probably not), you aren't in any danger of your water being corpse-contaminated.

That doesn't mean humans will ever overcome our fear of

dead bodies around water. Take the new process called aqua-mation. You already know about cremation, which uses flames to burn the flesh and organic matter, leaving only a skeleton. Aquamation uses water and potassium hydroxide to dissolve the dead body down to its skeleton. The aquamation process is better for the environment and doesn't use natural gas, a valuable resource. But the idea that a body can be dissolved in water drives some people wild with fear—especially when they find out that the water used in the process, which is not danger-ous in any way, is sent into the sewer system. Newspapers run headlines like "Have a Glass of Granddad!" With the subtitle: "Plan to Flush the Dead Down the Drain." That is a real head-line. Even worse, it appeared in a major, respected newspaper. Sigh. Don't drink Granddad, kids.

I went to the show where dead bodies with no skin play soccer. Can we do that with my body?

Say no more. If it's a fleshless cadaver playing soccer, you're definitely talking about the exhibition Body Worlds. The original Body Worlds, which is a traveling show, opened in Tokyo in 1995, and began touring the United States in 2004. (Keep an eye out, the merry band of corpses may be on their way to your town!) Millions of people have seen these exhibits. Some people absolutely love them and think the exhibits teach us about science, anatomy, and death. Other people call them "a gruesome Brechtian parody of capitalist excess." (Yeah, I don't know what that means either, but it sounds bad.) Either way, once you see the pregnant woman complete with cross-section fetus, or a man and woman having sex, or the flayed corpse playing soccer, it's hard to stop wondering about these strange, plastic bodies.

First of all: yes, they're real human corpses. And, with a few important exceptions, they wanted to be there on display. Around 18,000 people, mostly Germans, have added themselves to a list of body donors to Body Worlds. There's even a donation card at the end of the exhibition that you can fill out. One woman requested that her body be posed diving for a volleyball. All the bodies on display are made anonymous, so no one can go looking for the body of a specific person, like, "Is that Jake's corpse playing air guitar?"

Body Worlds is far from the first time humans have prepared corpses for long-term preservation and display. Like cooking and sports and storytelling and gossip, preserving corpses is a near  universal human pastime. From China, to Egypt, to Mesopotamia, to the Atacama Desert in Peru, people with special knowledge produced mummies using herbs, tar, plant oils, and other natural products, along with techniques like removing organs and hollowing out body cavities. Preservation became more precise during the Renaissance, when people figured out that you could inject fluids directly into a corpse's veins and the body's circulatory system would carry them to all the corpse's nooks and crannies. Ink, mercury, wine, turpentine, camphor, vermilion, and "Prussian blue" (ferric hexacyanoferrate) were just a few of the compounds used.

This brings us to plastination, the preservation technique used by Body Worlds. Plastination was originally developed for making anatomical specimens for students. But, with artistic finesse, it can also turn a dead body into a kind of weird plastic sculpture.

If you choose to donate your body and be plastinated, you'll be preserved with formaldehyde, dissected, and dehydrated. Your fluids and your squishy parts (water and fat) get sucked out when your body is dunked in a freezing bath of acetone, which you may know as the main chemical in nail polish remover. The acetone takes the place of water and fat in your body's cells. Remember how your body contains around 60 percent water? Now it's around 60 percent nail polish remover.

In the most important step, your acetone-filled body is then boiled in another bath, this time a bath of molten plastics like silicone and polyester, inside a vacuum-sealed chamber. The vacuum forces the acetone to boil and evaporate out of your cells. Then the molten plastic floods in. Now, with a little hands-on help from the living, your plastic-pumped corpse can strike a pose.

Depending on the type and amount of matter to be hardened, ultraviolet light, gas, or heat are used to solidify the posed corpse. Voilà! You have become a hard, dry, odorless corpse, frozen mid-volleyball save. The plastination of your full body can take up to a year and cost as much as fifty thousand dollars.

Gunther von Hagens, the German showman who pioneered the craft of keeping these corpses frozen in time, calls himself "the plastinator," which has a sort of professional wrestler or B-horror movie vibe to it. He runs the Institute for Plastination in Germany, where visitors can check out some of the fruits of his labors. But von Hagens has also had some bumpy moments in his career, which you should know about if you're considering donating your body to be plastinated for his traveling show.

Von Hagens was accused of profiting from illegal body trafficking: buying cadavers from hospitals in China and Kyrgyzstan that had no right to sell them. The people who died certainly didn't know their bodies were going to be posed playing the saxophone or holding their flayed skin for all eternity. It's too bad that Body Worlds started out with this reputation, because it turns out there are plenty of people who are happy to donate their bodies to the exhibition.

And don't confuse Body Worlds with BODIES . . . The Exhibition, a Body Worlds spinoff. This separate organization's website says that it displays "human remains of Chinese citizens or residents which were originally received by the Chinese Bureau of Police," including body parts, organs, fetuses, and embryos from the same source. The organization relies "solely on the representations of its Chinese partners," it says, and "cannot independently verify that [the remains] do not belong to persons executed while incarcerated in Chinese prisons." Oh. Executed prisoners. That sounds like a fun family activity.

So, if you attend one of these exhibits (or any display of human anatomical specimens that can't give you information about its sources), you may be looking at the remains of someone who wanted to display their body, and gave it up willingly and legally. But it's just as likely that the person would have been horrified that their body ended up like this.

A final tidbit to bear in mind about displays of human remains displays is that body parts occasionally go missing. In 2005, two mystery women stole a plastinated fetus from the Los Angeles Body Worlds exhibit. And in 2018, a man in New Zealand briefly made off with a couple of plastinated toes. Each toe was valued at more than three thousand dollars—pretty pricey toes, though not quite an arm and a leg.

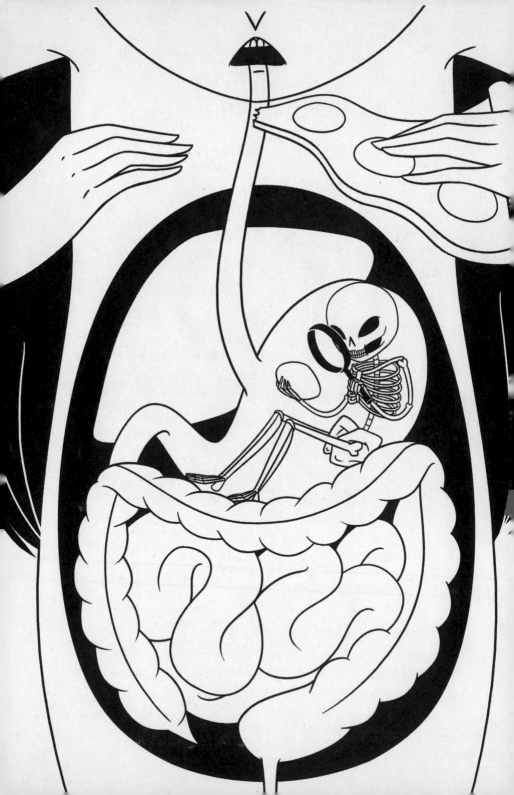

If someone is eating something when they die, does their body digest that food?

You're dead, but does your pizza go on?

Well, not really.

Food in your stomach doesn't stop being digested at the exact moment of death, but the process does slow down.

Here's the scene: you've just watched some internet videos, eaten a delicious slice of pizza, had a heart attack, and keeled over dead. In some ways, the pizza is already on its way to digestion. When you chewed the slice, you not only mechanically mashed up the pizza, you also mixed in digestive enzymes from your spit, which start to break down that sauce, crust, and cheese. Then you swallowed, your esophagus contracting, which sent that yummy ball of enzyme cheese into your stomach.

If you were still living, your stomach would be working to digest the food, secreting hydrochloric acid to break it down while muscular mechanical action was mixing and mashing it. But you're dead. Your stomach is no longer secreting and mashing, so the only things helping break down the pizza are the digestive juices remaining from *before* you died, and the bacteria present in your digestive tract.

So, let's say that they don't find your dead body for several days. Darn, this hypothetical pizza example is getting *dark*. Sorry. The medical examiner performs an autopsy on you, trying to determine when and how you died. When they open your stomach, that slice of pizza is going to become the forensic's best friend. Here's how.

If we know you ordered pizza at around 7:30 pm on Tuesday, and your corpse is found on Friday, the state and position of the partially digested pizza inside your body can give clues as to how long you were alive after eating. If there is a pile of barely digested pizza in your stomach, we know that you died soon after your last meal (which you did). If the pizza had been turned into a paste and sent on its merry way through the gastrointestinal tract, we'd know that you had time to digest, and died much later in the evening. This is all part of discovering the postmortem interval, a phrase that means "how long you've been dead."

Now, to be clear, "How's that pizza in the stomach look?" does not always provide a scientifically useful answer. Forensic pathologists do look at stomach contents for rough estimates, but there are other factors at play that affect digestion, such as medications, diabetes, how liquid the meal was, etc. Doctors look at food left in your stomach, finding everything from undigested gum (more common than you'd think) to bezoars, solid masses of accumulated indigestible material (protect yourself by not Googling it). But pathologists also have to examine your bowels. This process is much harder than opening the stomach, and much grosser. The pathologist will remove your intestines (which are nearly as long as a bus), place them in the sink, and slice down their whole length. My pathologist friend calls this "running the bowels." Then they sift through the gruesome tube. What's in there? Mashed up pizza remains, poop, medical anomalies? Who knows, that's part of the adventure. (An adventure that, again, makes me glad to be a mortician rather than a forensic pathologist.)

Keep in mind that if investigators don't have that receipt saying that you had pizza delivered at 7:30 pm, the undigested pizza won't be of much help. I myself ate leftover pizza at 10 am this morning. And again at 3 pm. I might have another slice now. (I don't have to explain myself to you.) But investigators would have no way of knowing when I ate that pizza. So the state of the pizza in my stomach wouldn't help them figure out when I died.

A bunch of undigested pizza sitting in your stomach might be useful in determining your time of death, but it's a big problem for an embalmer preparing your body for a family viewing. A whole pizza in your stomach means food hanging out, rotting, destroying the preservation vibes that embalmers are trying to achieve. That's one of the reasons why they use a tool called the trocar. A trocar is a large, long needle that an embalmer will poke into your abdomen, just below your belly button. The idea is to stick it in there, puncture your lungs, stomach, and abdomen, and suck out whatever's inside. This includes gas, fluids, poop, and yes, your pizza juices.

Maybe you don't want that undigested food sucked out with a trocar, because you hope someday, far in the future, it will be used to determine what people of your era ate. Take Ötzi, the 5,300-year-old mummy found by two German hikers on the border between Austria and Italy. When scientists examined the contents of his full stomach, they discovered what would have been Ötzi's last meal before his death by arrow in the back—murder most foul! Spoiler: the meal wasn't pizza. It was meat (ibex and red deer), einkorn wheat, and "traces of toxic bracken" (a type of fern). His diet was much higher in fat than scientists had expected it to be (relatable!). Because he didn't have time to digest, Ötzi's stomach was able to teach us wildly valuable things about life and diet 5,300 years ago. Perhaps someday your stuffed crust pizza and Flamin' Hot Cheetos will do the same.

Can everybody fit in a casket? What if they're really tall?

Listen, sometimes people just don't fit inside a casket. And funeral directors have to do something about it. It's our job. The family is counting on us. If we are left with no other options, *we will have to amputate their legs below the knee to make them fit.*

No! What the heck? *We don't do that.* Why does everyone think that's what funeral homes do to tall people?

Sadly, this whole amputation rumor is not just an urban legend. In 2009, it actually happened in South Carolina. Our story begins with the death of a man who was six feet seven inches tall. Which is tall, but not *that* tall by casket standards (more on that later). His body was taken to Cave Funeral Home.

Here's where things go from zero to "holy crap that's macabre" very quick. The funeral home owner's father would often do odd jobs around the funeral home, like clean and dress the bodies and place them in the casket. According to dear old dad, one day he made the executive decision to *sever the gentleman's legs at the calf with an electric saw and place the legs alongside him in the casket.* They even went ahead and held a viewing where only the man's head and torso were visible—for

obvious reasons. It wasn't until four years later, when a former employee came forward to reveal what he knew, that the man's casket was exhumed. Surprise! There were his legs, still tucked up beside him.

Everything about the decision to saw someone's legs off is baffling. I didn't believe this story when I first heard it, because cutting off a corpse's feet or legs is something no funeral employee would do. It goes against common sense and professional ethics. Even if the dead man's wife was begging, "Please, cut his legs off to bring me peace of mind," doing so would break—you guessed it—abuse of a corpse laws, which are designed to protect a corpse from being maimed. It would also be a complete mess. Not that the mess is the biggest concern here, but it's worth mentioning.

Honestly, the hardest part of this story to reconcile is the idea that the corpse wouldn't fit in a casket. Six feet seven inches isn't that absurdly tall when it comes to caskets. Most average-sized US caskets can accommodate someone who is six feet seven, or even seven feet tall. Even if the funeral home only had caskets in stock that were on the shorter side, they could easily order a larger casket, or even remove some of the inner lining of an in-stock casket to make some leg room. It's hard to conceive of there ever being a time when sawing a guy's legs off seemed like the more sensible option.

Okay, but what if the dead person is seriously tall, like Manute Bol, one of the two tallest men ever to play professional basketball in the NBA. Bol towered over . . . well . . . pretty much everyone, at seven feet seven inches, and his "wingspan" (measurement from fingertip to fingertip) was an unprecedented eight feet six inches. Can people that tall have caskets?

For the record, anybody can have a casket. "Oversize" caskets do cost more. I'm not saying this extra cost is fair, but it's the reality of how the funeral industry operates. I've heard of caskets as long as eight feet. Even a quick internet search will turn up companies that specialize in making caskets for people who are larger than what's considered standard size.

Finding a company to build a casket for someone who is seven feet seven inches tall might be harder—but there are custom casket companies who will build to your exact specifications. I can't think of a realistic scenario where an extra-wide or extra-tall casket couldn't be manufactured to fit any size corpse. Heck, there are even downloadable plans online for making your own DIY casket. You feelin' crafty?

Of course, if you're taking an extremely tall person to be buried, you might run into some further problems at the cemetery. If our pal Manute wanted to be buried in a conventional cemetery—with manicured lawns and rows of tidy graves—he'd have needed to ask about plot size. Each cemetery plot has set dimensions, usually for an "average-sized" person. When someone is buried in that plot, their casket is put inside a grave liner or vault—a concrete container that keeps the ground level. That grave liner is usually of an "average" size as well. If someone is extremely tall, they may not fit, and more than one plot (and maybe a custom vault) will have to be purchased.

This all sounds frustrating. But people who stand seven feet seven inches tall have lived their whole lives dealing with the fact that they almost never fit society's definition of "standard" and "average." They've struggled to find the right size shoes, shower heads, door frames, jeans—just about everything. An extra-large casket and burial plot are just two more things they'll have to have custom-made.

They might decide to skip the custom orders and go for a natural burial, straight into the ground in an unbleached cotton shroud. That's perhaps the easiest option of all. The cemetery can even dig a longer hole for a grave—no casket or vault required!

But what about cremation? From my experience working in a crematory, and talking with other crematory workers, cremating an extremely tall body shouldn't be a problem. Most modern cremation chambers can handle a body that is around seven feet tall, and you wouldn't run into problems unless the body was *almost nine feet tall*. Theoretically, such a retort could even handle the body of Robert Wadlow, the tallest recorded person who ever lived. Robert was eight feet eleven inches tall. He wasn't cremated, but you bet he had a custom casket. It was reportedly over ten feet long and weighed over 800 pounds.

If you are approaching seven feet tall, I would recommend researching caskets and burial plots *before you die* (not after). Talk with family and friends about how to communicate with a funeral director. Say to them, "Tell the funeral home I'm six feet ten inches and 415 pounds, just so there are no surprises." This can empower your family to advocate for your corpse if anybody gives them trouble.

If your funeral director acts like they don't know how to handle a very tall person and doesn't know about custom caskets, you might want to double-check that they aren't actually eight chihuahuas standing atop one another in a trench coat. Morticians can handle almost anything. There is always a way that doesn't involve the creative use of an electric saw.

Can someone donate blood after they die?

Blood is strongly associated with life, so I don't think anyone's first choice is a transfusion of stagnant blood from a corpse. But blood beggars can't be blood choosers and donating blood after you die is safer and more effective than you'd think.

In 1928, Soviet surgeon V. N. Shamov decided to investigate whether blood from a dead body could be used to keep a living one from the same fate. He started his experiments with dogs. As with most animal testing, the design of the experiment sounds a lot like—how do I put this?—torture.

Shamov and his team removed 70 percent of a living dog's circulating blood volume. In other words, they took out nearly three-quarters of all the blood in the dog's body. Then the team washed out the depleted bloodstream with warm saline, to bring the total level of exsanguination (a cool word that means the draining of blood) to 90 percent, a lethal level.

But hope was not lost for this brave laboratory pupkins. Another dog had been killed just hours earlier. The dead dog's blood was infused into the dying dog and, as if by magic, the dying dog came back to life. Further experiments demonstrated that so long as the dead dog's blood was removed within six hours after death, the living recipients of the blood did just fine.

From here, blood donation gets a little less *Saw* and a little more *Frankenstein*. Two years later, the same Soviet team successfully tested cadaver blood donation on humans, and spent the better part of the next thirty years happily transfusing the life-giving fluid from the dead to the living. In 1961, Jack Kevorkian, who later earned the nickname "Dr. Death" for helping patients who wished for medical assistance in dying, became the first American doctor to attempt the practice.

These experiments help to prove that dying isn't like switching off a light. Just because a person has died—they've stopped breathing, and their brain shows no electrical activity (as we discussed in the coma/brain death question)—does not mean that their body has suddenly become useless. As Dr. Shamov wrote, "The corpse should no longer be considered dead in the first hours after death." A heart kept on ice can be transplanted up to four hours after death. A liver, ten. A particularly good kidney will last twenty-four hours, and sometimes as long as seventy-two if doctors use the right equipment after surgery. This is known as the "cold ischemic time." Consider it the five-second rule, but for organs.

As long as the death was relatively sudden and the dead person was in otherwise good health, cadaver blood remains usable, as Dr. Shamov discovered, for up to six hours. In other words, donation is a go—though obviously it is better if the blood isn't tainted with medication or communicable diseases. White blood cells have several days of activity left in them after the heart stops beating. If the blood is sterile and in good condition, cadaver blood donation is perfectly fine.

So if these transfusions are possible, why aren't they popular? A few reasons. Cadaver blood donation, let's be honest, is sort of a one-time thing. Doctors realized early on that living donors can give blood (and score free cookies) many times a year—as often as every eight weeks. While there are limited numbers of healthy, disease-free cadavers to exsanguinate, we can pro-mote blood donation through blood drives; donation centers can welcome back repeat (living) customers for years on end.

Blood from living donors also avoids the ethical implications of giving someone corpse blood without their knowledge. If you get a pair of lungs from an organ donor, there is an obvious knowledge of their origins (psst, from a dead person). A patient in the midst of a crisis might need the blood too badly, and be too unconscious, to stop and have an informed chat about how their donor blood came gushing out of a dead man's neck.

Speaking of gushing out of the neck, that is actually sort of what happens. Without a beating heart to pump the blood out, cadaveric blood donation requires gravity to do the work. If pathologists need to get blood out of a cadaver, the simple option is to open a large vein in the neck and then tip the head down. Embalmers at your local funeral home have a more sophisticated drainage sys-tem, so gravity is not required. As embalm-ing fluid is pushed into the body, blood is pushed out, rolling down the table and into the sewer system. When I get the call from my local blood center asking for donations, I think of all the blood from embalming pro-cedures just pouring down the drain.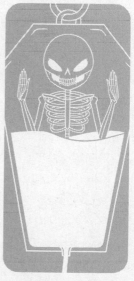

The most striking reason cadaver blood donation is not done is the stigma of blood from a corpse. Strange, since corpse parts are used in medicine all the time. I found out that one of my friends has tissue from a cadaver butt in her mouth. Turns out quite a few people do. When gums are receding, due to teeth grinding or health issues, they can be rebuilt by implanting cells from the butt of a human cadaver. So, cadaver butt is in, but cadaver blood is out.

I reached out to the Red Cross to get their official policy on cadaver blood donation but, as of this writing, they have yet to respond.

We eat dead chickens, why not dead people?

It is my sincere belief that you are never too young to ask the hard questions about cannibalism. So, let's dig in* to the topic of eating human flesh!

You might assume the answer is obvious: "We don't eat dead people because it's horrifying! Morally repugnant!" Not so fast there. Eating a dead human may be horrifying *to you*, but humans throughout history have practiced mortuary cannibalism. Mortuary cannibalism is when the relatives, neighbors, or community members consume the flesh, or ashes, or both, of a dead person. Imagine that after Aunt Chloe died you sat around a fire eating little roasted bits of her and it was all totally normal.

Without judging other cultures for their cannibalism, we can agree that eating human is a big ol' taboo in the twenty-first-century developed world. We consider it morally wrong, something practiced only by the most diabolical serial killers and Donner Party members.

Beyond the taboo, there are more practical reasons for not eating other humans. One, human meat is hard to get, and two, human flesh just isn't that nutritious or good for you.

Let's tackle the "hard to get" problem first. Someone would

* Wink.

need to die for you to feast. Even if that person died of natural causes, you're not legally allowed to lay claim to a dead person just because they look tasty.

What laws would you be breaking if you did get a dead human to eat? Amazing fact: *cannibalism is not against the law.* It's not criminal to eat human flesh, but acquiring the human flesh (even if the dead person wanted you to eat them) is breaking the law. The laws you're breaking are . . . wait for it. . . . remember these? Welcome back, abuse of corpse laws! It's considered desecration and mutilation to eat a dead body. You might also be charged with stealing the corpse. Stealing is bad, right? The dead guy's mom wanted to bury him in the family plot but now he's missing a leg, for heaven's sake.

But let's say, by some hypothetical scenario, it wasn't illegal for you to desecrate the dead by eating them. Is human meat a healthy choice?

No.

In 1945 and 1956, two researchers analyzed the donated bodies of four adult males, and estimated that the average male offers about 125,822 calories from protein and fat. That number is far below what other red meats like beef or boar can offer.

(Yes, you heard me, humans are red meat.)

That's not to say those precious calories wouldn't be helpful in a life-or-death starvation situation. In 1972, Pedro Algorta's plane crashed in the Andes mountains. Some didn't survive the crash. Pedro, starving, began to eat the dead people's hands, thighs, and arms. Human meat wasn't ideal, but this is a seventy-one-day ordeal of starvation we're talking about. Pedro said, "I always had a hand or something in my pocket, and when I could, I would begin to eat, to put something in my mouth, to feel that I was getting nourished." In that extreme scenario,

Pedro didn't care that human meat isn't the best source of calories and protein. He just wanted to live.

Evidence suggests that humans have never thought eating other humans was a great option, nutrition-wise. An archaeologist at the University of Brighton, in England, found that early species of humans, like Neanderthals or *Homo erectus*, had cannibalistic tendencies. But if they ate their own kind, it was for ritual purposes, not dietary purposes. Again, humans just don't provide enough calories to compete with something like a mammoth, which would have provided a (totally worth it) 3.6 million calories. In addition, almost half the calories in a human come from fat. Humans aren't even a heart-healthy option! We're all-around bad eating.

And, when considering the pros and cons of eating humans, you also have to think about disease. I know, you're thinking, "Caitlin! Haven't you said like, a thousand times, that dead bodies aren't dangerous? That a dead body isn't going to give me a disease? What's the deal!"

Yes, those statements are still true. It is unlikely that a dead body is going to give you the same disease that killed the person—or any other disease, for that matter. Most pathogens, even nasty ones causing tuberculosis or malaria, just don't live that long in a corpse after death. But keep in mind *I never told you to eat the corpse.*

Your question mentioned eating dead chickens, so let's say you live on a farm. You walk out to the yard on a hot summer day to feed your hens and discover Big Bertha has bit the dust overnight. You notice that although Big Bertha is not visibly decomposing yet, she has some flies buzzing around her. She's starting to bloat. What did she die of? Jeez, is that a maggot?

Now ask yourself: are you hungry? Probably not.

Humans in the developed world prefer our meat maggot-, ailment-, and bloat-free. (Not always, though. There are cultures that consider putrid meat a delicacy. My favorite example is hákarl or fermented shark, which is a beloved national dish in Iceland. The shark is buried, fermented, and hung to dry for months until its debut as a pungent, rotten treat.)

More common meat, like the cows and chickens at the grocery store, have been killed specifically for human consumption. Once the animal has been slaughtered, the meat is immediately cleaned and stored in a refrigerator or curing shed to avoid the bacterial growth and autolysis that make meat decompose, turn gross colors, and smell weird. The chicken, cow, or pig meat you buy at the store or butcher shop was not found lying dead somewhere. There are about a billion laws that stop "Roadkill Ranchers" from selling carrion to the public.

Humans don't do well eating rotten meat—or diseased meat, for that matter. We greatly prefer to eat fresh, healthy meat. But very few wholesome, strapping, good-for-grilling people just up and die. Most dead people have health issues that would make them at best unappetizing, at worst unsafe to consume. Also, think of it this way. Even if the animal you ate had some sort of disease, most diseases are not zoonotic. That is to say, a human can't get an animal disease by eating that animal. (Ebola is one of the rare exceptions.)

But if you're going to eat a *human* corpse it's a different story. It is possible to contract blood-borne viruses such hepatitis B or HIV. Unlike when you eat animals, if you eat diseased human flesh, you could end up suffering from the same illness.

"No problem," you might say. "I'll just cook the human meat well-done, and it'll be good eatin'!"

Think again.

Humans can have abnormal proteins called prions. These proteins have lost their shape and proper function, and infect other, normal proteins. Unlike a virus or an infection, prions don't have DNA or RNA, so they can't be killed by heat or radiation. They're tough little suckers that like to hang out in the brain and spinal column, spreading lesions and chaos.

When talking about prions, scientists often point to the Fore people of Papua New Guinea. As late as the 1950s, anthropologists documented an epidemic of a neurological disease called kuru that was killing members of the tribe. Kuru is a disease caused by prions in the brain. The spread of kuru was traced to the tribe's ritual of eating of human brains after death. The infected suffer from muscle spasms, dementia, and uncontrollable laughing or crying. The end result is a brain literally full of holes—and then death.

After a member of the Fore tribe died, their family would eat the prion-filled brain and the disease would spread, sometimes lying dormant in an infected person for up to fifty years. It wasn't until the Fore ended the practice of brain-eating in the mid-twentieth century that kuru began to decline.

To return to my initial point, lovingly caring for a corpse that died from kuru won't hurt you. But *eating it* will.

I think we've shown that abuse of corpse laws, low nutritional value, and infectious diseases are pretty solid reasons to say, "Maybe just . . . don't eat people?" We may get to a place one day where lab-grown human is on your local restaurant menu (yes, someone is already developing this technology), but until then I think it's best to stay away from the *other* red meat.

What happens when a cemetery is full of bodies and you can't add any more?

If you've got more bodies than you know what to do with, the first sensible option is to expand. Expanding could mean adding more land to the current cemetery (making room for more graves), or opening an entirely new cemetery nearby.

"But this is a big city!" you say. "We don't have any extra green fields lying around for dead people!" Okay, then what about expanding . . . up? That's right—cemeteries are going vertical. After all, city dwellers live in skyscrapers and apartments, stacked on top of one another. But when we die, everyone is supposed to be buried spread apart on hundreds of acres of rolling land? One architect who designs these multistory cemeteries said, "If we have already agreed to live one on top of the other, then we can die one on top of the other." Touché.

The Yarkon Cemetery in Israel has started adding burial towers that will ultimately hold 250,000 graves. The towers even respect Jewish custom by filling the burial columns with dirt so that the graves are connected to the earth. At the moment, the world's tallest cemetery is in Brazil. Memorial Necrópole Ecumênica III contains thirty-two stories of graves, and also

has a restaurant, concert hall, and gardens filled with exotic birds. When I was in Tokyo, Japan, I visited a multistory building that houses thousands of cremated remains (delivered to personal visiting rooms by automatic conveyors that locate and  fetch the correct urn). It looks like a typical office building, blending into the city around it, and is located right off the subway stop for convenience. There are more vertical cemeteries planned in places as diverse as Paris, Mexico City, and Mumbai.

Think of it this way: even a generic, sprawling cemetery goes vertical when they add mausoleums to the property. Mausoleums are those squat buildings in the middle of the cemetery where people are buried in cubbies in the wall, called crypts. If you buried people in single graves in the ground, you could run out of space quickly. Build a mausoleum, and that single grave space turns into a stack of three or four (or more) cubbies, on top of one another. Cemeteries advertise different crypts depending on their height from the ground, with names like "heart level" and "sky level." The crypts closest to the floor are called "prayer level," as in, easy to kneel and pray in front of. (I guess "ground level" wasn't as marketable.)

If you don't want to build upward to house more bodies, another option is to recycle the graves that are already there. This may sound horrifying to you, if you're used to the idea of Grandpa's gravesite belonging to him forever. In Germany and Belgium, public graves are offered for a set time limit, somewhere between fifteen and thirty years, depending on the city.

When your time is up, your family is contacted and given the option of continuing to pay to rent the grave. If they can't or don't want to pay, your body will either be moved deeper down into the ground (to make room for new friends), or relocated to a communal grave (*lots* of new friends). In these countries, you rent a grave; you don't own one.

Why is America different? Why do we pay for something called "perpetual care," believing that the cemetery is going to take care of our grave *forevvvvvver*? The idea of a forever-and-ever grave started because America was just so large. In the nineteenth century, burial moved from overcrowded (read: smelly) urban graveyards to sprawling rural cemeteries. These rural cemeteries hosted picnics, poetry readings, carriage races. They were the places to see and be seen. The thought was that with the size of the country, we could keep burying people forever. Everyone gets a grave!

Not so fast. In the twenty-first century, the death rate in the United States is 2,712,630 people per year. That works out to a little over 300 deaths per hour. Or, five people per minute. But, even with that much death, the looming crisis of burial space is misleading: America still has tons of space for graves. It's finding burial space near cities and near their already buried loved ones that makes things more complicated. For that reason, New York City has to solve this problem more urgently than North Dakota does.

Now, some countries that complain about lack of burial space *really mean it*. Good examples would be Singapore and Hong Kong, the third and fourth most densely populated countries in the world. In Singapore, for every square mile there are more than 18,000 people. Every. Square. Mile. 18,000. People. In the United States, for every square mile there are only 92 people. Womp womp. Sorry, U.S., when Singapore clutches its pearls and says, "We have no land to bury our dead," they're serious. Take Chua Chu Kang cemetery in Singapore, which is the only cemetery *in the whole country* still open for burial. Singapore is so small, geographically, that there is no affordable open land to create more cemeteries. The government passed a law in 1998 that says a person can only be buried there for fifteen years. When your fifteen years is up, your body is dug up, cremated, and stored in a columbarium (a building like a mausoleum, except for cremated remains).

If you're willing to walk away from burial altogether, cremation and alkaline hydrolysis (remember—cremation with water instead of fire) are two excellent options. You end up with about four to six pounds of ashes, which can be scattered or placed on the mantel. But if it's burial you desire, perhaps it's time to join the rest of the world and—gasp!—recycle our graves. Once Grandma has had her time to decompose, her bones need to step aside for a whole new generation of rotting corpses. I wonder if anyone has ever written that exact sentence before? I wonder that a lot.

Is it true people see a white light as they're dying?

Yes, they do. That glowing white light is a tunnel to the angels in heaven. Thanks for your question!

Truth is, I don't have a perfect explanation for why some people see a white light when they are near death. In fact, *no one* has a perfect explanation yet. Religious folks may see the light as a supernatural gateway to the afterlife; scientists may see the light as caused by oxygen deprivation in the brain.

What we do know is that these strange experiences *are* happening; there are just too many reports from across the religious and cultural spectrum for them not to be real. People who have survived traumatic life-threatening situations share a set of eerily similar experiences, which scientists refer to as near-death experiences, or NDEs. Spooky as they may seem, near-death experiences aren't even particularly rare. Approximately 3 percent of Americans say that they've had one. That number was even higher (18 percent) in a study done with elderly hospital patients.

It's important to remember that not all near-death experiences are created equal. Not everyone finds themselves walking into a sparkling white light while scenes of childhood pets and awkward job interviews pass before their eyes. In one study, about half of the people who had a near-death experience said they were fully aware they were dead (which could be good or bad, depending on how chill with death you are). One in four people said they had an out-of-body experience. Only one in three actually moved through the good ol' tunnel. Also, some bad news: we imagine NDEs as being positive and blissful, but that was only true about half the time. Turns out they can be pretty terrifying, too.

There are scholars who believe near-death experiences have been happening in cultures throughout human history: ancient Egypt, ancient China, medieval Europe. These cultures (and countless more) have tales of religious experiences that match up almost exactly with near-death experiences. This brings up an interesting chicken-and-egg dilemma. Are near-death experiences some kind of universal religious experience? Or are religious experiences caused by the action of the human brain, basic neuroscience and biology?

The setting—the vibe, if you will—of an individual's NDE can also be determined by the society in which they live. For example, American Christians might have angels greeting them in the tunnel, while Hindus might have someone sent by the god of death. Gregory Shushan, a researcher at Oxford University, writes about wildly different accounts of NDEs, with the cast of characters drawn from the person's culture: "I remember one describing Jesus in the form of a centaur riding a chariot; and a man whose heart was beating on the outside of his chest, and with hair in the shape of a bishop's hat."

What makes it even harder for scientists studying NDEs is that you don't have to be near death to have a near-death experience. Researchers at the University of Virginia found that just over half of patients who recounted having an NDE were not actually in medical danger. Death, as it turns out, was not so near after all.

So let's talk some potential (scientific) explanations for why this might be happening. If you're a brain doctor, you're likely to explain NDEs using some fancy and confusing language like "disturbed bodily multisensory integration." Other explanations include endorphins released in the brain, too much carbon dioxide in the patient's blood, or increased temporal lobe activity.

But let's go for an even simpler explanation, and look at another group of people who experience the eerie tunnel of light: fighter pilots. Flying at high speeds can cause something called a hypotensive syncope, which happens when there isn't enough blood and oxygen getting to the brain. When this occurs, the pilot's vision starts to go, with the edges going first—creating the experience of looking down a bright tunnel. Sound familiar?

Scientists believe that seeing this light at the end of the tunnel is the result of retinal ischemia, which happens when there isn't enough blood reaching the eye. As less blood flows to the eyes, vision is reduced. Being in a state of extreme fear can also cause retinal ischemia. Both fear and decrease in oxygen are associated with dying. In this context, the extreme white tunnel vision characteristic of NDEs starts to make much more sense.

If you are religious, you may believe that God (or gods) are capable of magical things. But scientists (even the ones who believe in God) also believe that the brain is also capable of making things seem and feel magical. They believe that biology is what shapes our final moments. I'm not personally religious, but I am 100 percent game for centaur Jesus riding a chariot coming to pick me up for my descent into death.

Why don't bugs
eat people's bones?

It's a lovely summer day and you're having lunch in the park. You bite into a fried chicken wing, munching on the crispy skin and juicy flesh. Is your next move cracking into the bones, crunching them like the giant in "Jack and the Beanstalk"? Probably not.

If you yourself wouldn't eat a pile of animal bones, why would you expect a beetle to show up and eat your bones? We expect too much from necrophages, the unsung heroes of the natural world. They are the death eaters, the organisms that fuel up by consuming dead and rotting things—and bless their hearts! Imagine, for a moment, what the world would look like without the assistance of the consumers of dead flesh. Corpses and carcasses everywhere. That road kill? It's not going anywhere without the help of necrophages.

Necrophages do *such* a good job getting rid of dead things that we expect them to perform miracles. It's like how if you do *too* good a job of cleaning your room, then your mom will expect perfection every time. Better to not set expectations so high. It's just not worth the risk.

The corpse-nosher ranks are filled with diverse species. You have vultures, swooping down for a roadside snack. You have blowflies, which can smell death from up to ten miles away. You have carrion beetles, which devour dried muscle. A dead human body is a wonderland of ecological niches, offering a wide range of homes and snacks for those inclined to eat. There are plenty of seats at death's dinner table.

Remember the dermestid beetle? The helpful cuties we'd enlist to clean your parents' skulls? Their job is to eat all the flesh off *without* damaging the bone. Let's be clear: *we don't want them to eat the bone.* Especially because other methods of flesh removal (like harsh chemicals) will not only hurt the bones, but might damage certain types of evidence, like marks on bones, which could be useful in criminal investigations. That's why you bring in a colony of thousands of dermestids to do the dirty work. Plus, while you were over here complaining that they don't eat enough bones, the beetles were also eating skin, hair, and feathers!

All right, but to your question: why don't they eat bones, too? The simple answer is that eating bones is hard work. Not only that, but bones are not nutritionally useful to insects. Bones are mostly made of calcium, something insects just don't need a lot of. Since they don't need much calcium, insects like dermestids haven't evolved to consume it or desire it. They're about as interested in eating bones as you are.

But, here's a dramatic twist: just because these beetles don't usually eat bone doesn't mean they won't. It's a cost-reward thing. Bones are a frustrating meal, but a meal is a meal. Peter Coffey, an agriculture educator at the University of Maryland, told me how he learned this firsthand when he used *Dermestes maculatus* to clean the skeleton of a stillborn lamb. Adult sheep bones are robust, "but in fetuses and newborns there are several places where fusion is not yet complete." When he removed the lamb bones after the beetles finished cleaning them, "I noticed small round holes, about the diameter of a large larva." It turns out beetles will go after less dense, delicate bones (like those of the stillborn lamb), but, Peter says, "there has to be a perfect storm of good environmental conditions and poor food availability before they'll resort to bone, which would explain why it's not more commonly observed."

So, while dermestids and other flesh-eating bugs do not usually eat bone, if they get hungry enough, they will. Humans behave the same way. When Paris was under siege in the late sixteenth century, the city was starving. When people inside the city ran out of cats and dogs and rats to eat, they began disinterring bodies from the mass graves in the cemetery. They took the bones and ground them into flour to make what became known as Madame de Montpensier's bread. Bone appetit!  (Actually, maybe don't bone appetit, as many who ate the bone bread died themselves.)

It seems like no creature out there wants to eat bone, really prefers bone. But wait, I haven't introduced you to *Osedax*, or the bone worm. (I mean, it's right there in the name, people. *Osedax* means "bone eater" or "bone devourer" in Latin.) Bone worms start as tiny larvae, floating out in the vast blackness of the deep ocean. Suddenly, emerging from the void above is a big ol' dead creature, like a whale or an elephant seal. The bone worm attaches, and the feast begins. To be fair, even *Osedax* don't really devour the minerals in the bone. Instead, they burrow into the bone searching for collagen and lipids to eat. After the whale is gone, the worms die, but not before they release enough larvae to travel the currents waiting for another carcass to come along.

Bone worms aren't picky. You could throw a cow, or your dad (don't do that), overboard and they'd eat those bones, too. There is strong evidence that bone worms have been eating giant marine reptiles since the time of the dinosaurs. That means the whale eaters are older than whales themselves. *Osedax* are nature's peak bone eaters, and they're even sorta nice to look at, orangey-red floating tubes covering bones like a deep-sea shag carpet. Pretty amazing, given that scientist didn't even know these creatures existed until 2002. Who knows what else is out there in the world, devouring bone?

What happens when you want to bury someone but the ground is too frozen?

I was raised in Hawai'i, a place that's not known for its harsh winters. As an adult, I own a funeral home in California, a place . . . also not known for its winters. To sum up, I'm a terrible person to answer this question. I've never had to drive a jackhammer into the icy, frozen topsoil. The family and guests at our graveside burial services aren't bundled against the cold; they're fanning themselves and longing for the comfort of their air-conditioned cars.

But what about Canada? Norway? Places trapped in the depths of winter's icy embrace? Frozen ground is frozen. It's like rigor mortis in a dead body—much harder and more rigid than you'd expect. It's no easy task to drive a tool through the dirt and dig a grave. That's why, for most of human history, they just . . . didn't.

Back in 1800s America, if a person died during a harsh winter, they couldn't be buried until spring. To wait out the cold, the body would be put in what's called a receiving vault. A receiving vault was an outdoor structure that looked a lot like a

tomb. All the coffined corpses of folks who died at an inconveniently frigid time of year would go into this communal tomb. Since it was already cold as heck outside, the receiving vaults would act as natural refrigerators.

There were also plainer buildings for winter body storage that had even more "tell it like it is" names: dead houses. Dead houses were used in Europe, the Middle East, parts of the U.S., and Canada. They were also called mort houses or corpse houses. In the nineteenth and twentieth centuries, maybe even as early as the seventeenth century, people would put their dead in these dead houses to wait out the winter.

I may be a warm weather burier, but I happen to know an archaeologist, Robyn Lacy, who specializes in these dead houses. "Some of them still exist to this day," she tells me. "Not only do they exist, they're still in use!" In fact, you might walk by a

dead house on your stroll through the local cemetery. Just look for a simple wooden (sometimes brick) structure that could be mistaken for a tool shed.

For many years, a wintertime funeral procession ended not at the grave, but at the dead house. Normally, the mourners would head straight to the burial site, but if the ground was frozen the body had to wait for the spring thaw in the afterlife equivalent of the DMV.

Other cultures gave up on burial entirely. High in the mountains of Tibet, where the ground is often too rocky and frozen for burial, and where not enough trees grow to perform cremations, a different kind of death ritual developed. To this day, bodies are laid out in an open area for a sky burial, a lovely name for the dead body being consumed by vultures. Your cat *might* eat you after you die, but a vulture *can't wait* to rip you to pieces and carry you off into the sky.

But my home country, the United States, is perhaps not (yet) ready for burial by vulture. What is done with a body when the ground is frozen *now?* Thanks to technology, dead houses have fallen out of fashion (although I still use "dead house" as a nickname for my funeral home).

Most cemeteries in the U.S., even places with severe winters, can and will bury a corpse no matter how frozen the ground is. In some places, it's required by law. Wisconsin and New York both restrict cemeteries from storing bodies until warmer weather. They require that a cemetery bury a corpse within a reasonable time period, subzero temperatures or no.

On the other hand, there are still rural cemeteries that don't have the manpower or equipment to break through frozen ground. These rural areas may not even have access to the snowplows needed to allow the body to be driven on desolate, winter roads to the cemetery. In that case, they turn to good old refrigeration. The body will await the return of spring under refrigeration, either at a funeral home or sometimes at the cemetery itself.

There are pros and cons to using a body refrigeration unit until it is warmer outside. Cons are that in long winters corpses can pile up (not literally—just, like, there'll be a bunch of them stored in the fridge). Also, the longer the body is held in refrigeration, the higher the cost. On the pro side, unlike a receiving vault or dead house, there are no warm days in a refrigerator. No stinky surprises. Embalming can be used to slow decomposition for the unburied body as well.

But if the cemetery is able (or forced by law) to dig a grave through the frozen ground, there are generally two ways to go about it: break the ground or thaw it. Or a combination of both. Breaking the ground requires a construction jackhammer. This is not a quick process. It can take something like six hours just to blast four feet down into the frost. Another option is a backhoe outfitted with fearsome-looking "frost teeth." The frost teeth are metal arms, several feet long, attached to either side of the backhoe scoop. They look like backhoe fangs. Like a mechanical Dracula: "I vant to deeg your grave!" The fangs break the earth, allowing the backhoe to scoop out the frozen soil.

Instead of digging right in to the frozen ground, some cemeteries will attempt to thaw it first. There are a few ways to do this. Heated blankets can be laid over the future grave, which is pretty adorable. Lit charcoal can be spread over a future plot. There are also metal domes large enough to be placed over the grave, heated inside by propane. This setup sort of looks like there's a giant barbecue grill going in the middle of a cemetery. Not great for public relations, but you gotta do what you gotta do.

The only drawback with thawing the ground before breaking it is that you have to wait. The process takes between twelve and eighteen hours, even as long as twenty-four hours. But that's better than waiting the whole winter, right?

Don't worry if your grandpa's corpse needs to be buried when the ground is frozen. It might take a little longer, and he might have to hang out in cold storage for a bit, but he's going in the ground. Unfortunately, with all that extra work and/or corpse storage comes, you guessed it, extra costs. No such thing as a free corpsicle!

Can you describe the smell of a dead body?

Well, just how dead are we talking here?

If a person is newly deceased, they will smell very much like they did when they were alive. Did they drop dead suddenly, showered and perfumed? Then they'll smell showered and perfumed. Did they die after a long illness, in a musty hospital room? Then they'll smell like illness and musty hospital.

What the body *doesn't* do in the first hour or so after death is bloat, turn green, and burst out with maggots. I don't care how hot and humid it is outside, this is not a horror movie and that's not the timeline. We serve families at our funeral home who want to keep Mom's body at home but are worried about the "odors" of death. After explaining the not-going-to-be-swarming-with-maggots thing, we explain that they should start cooling Mom down with ice packs if they plan to keep her for more than twenty-four hours.

The reason dead bodies don't start smelling right away is because the classic "scent of decay" comes from decomposition, and decomposition emerges over several days. Remember, when a person dies, the bacteria in their intestines do not die with them. Not only do these gut bacteria not die, they are still hungry. *Hangry*. Ready to break down your body into organic material for other purposes.

It's not just hangry gut bacteria, either. The human body is teeming with life, a whole ecosystem of microbes. As they break down their brand-new food source—your dead body—the microbes give off gas made of VOCs, or volatile organic compounds. The prime stinkers here tend to be sulfur-containing compounds, which makes sense if you've ever experienced an especially potent and sulfuric eggy fart. Sulfur is the culprit in many a stink.

When specially trained cadaver dogs are searching the woods for a dead body, they're sniffing for VOCs. These smells also attract blowflies, which have scent receptors that lead them to the body. The sweet smell of decomp (a.k.a. *odor mortis*) tells them that the corpse over there is the perfect place to land and lay their eggs in the open orifices of the body. A short time later, fly larvae (maggots) are everywhere. Congrats, momma blowfly, on finding the perfect egg-laying locale.

Two of the best-known chemicals in dead body aroma are the aptly named putrescine and cadaverine (after "putrid" and "cadaver"). Scientists believe these foul smells are acting as necromones, that is, chemicals that trigger attraction or avoidance around dead things. If you are a cadaver dog or blowfly, these smells tell you you've found the dead body you're looking for. If you are a scavenger animal that eats carrion (decayed animal), these necromones will smell like a delicious lunch. If you are a boring old human—say, a funeral director—the smell will strongly encourage you to leave the room in search of fresh air.

Most bodies that come into the funeral home aren't in full-on decomposition mode. They haven't had time to get there. To prevent them from getting there under our watch, we put them straight into refrigeration, which slows decomposition way down. But that doesn't mean we don't get "decomps"—the industry term for bodies that haven't been found for several days, or even several weeks.

Those who have smelled a decomposed body rarely forget the experience. I conducted an informal poll of funeral directors and medical examiners, asking them to tell me how they describe the unforgettable smell. They offered thoughts ranging from "like roadkill, but way bigger," to "like rotting vegetables—soggy brussels sprouts or broccoli," to "rotten beef trapped in your refrigerator." Other examples: "rotting eggs," "licorice," "garbage can," "sewage."

And me? Oh, how to describe the smell of a decomposing human body—what poetry is needed! I get a sickly-sweet odor mixed with a strong rotting odor. Think: your grandma's heavy sweet perfume sprayed over a rotting fish. Put them together in a sealed plastic bag and leave them in the blazing sun for a few days. Then open the bag and put your nose in for a big whiff.

Though we may not have one single way of describing the smell of a decomposing human body, we do know that dead human smell is unique. Even though our untrained noses aren't likely to make that fine a distinction, researchers have found that humans have a "singular chemical cocktail," our own *eau de decomp*. Of the smelly compounds found in putrefaction gas, eight compounds give us humans our own special stink. Well, not 100 percent "our own" or "special," as pigs have those compounds, too. Darn it, pigs, can't we have something nice just for us?

Interestingly, humans were once much more accustomed to the stink of death, largely thanks to poor refrigeration and body preservation techniques. My longtime friend Dr. Lindsey

Fitzharris studies anatomy and dissection rooms from the nineteenth century. You think the refrigeration units at modern-day funeral homes smell bad? Boy howdy, be glad you weren't in a dissection room two hundred years ago. Medical students who performed the dissections, trying to learn more about the mysterious anatomy of humans, described "rancid corpses" and "putrid stenches." What's worse, the cadavers were stored in unrefrigerated rooms, stacked like piles of wood. Body handlers witnessed rats "in the corner gnawing bleeding vertebrae," and swarms of birds coming in to "fight for scraps." The young students might even have slept in a room directly next door.

In the mid-1800s, Dr. Ignaz Philipp Semmelweis noticed that new mothers who were treated by midwives fared much better than those who were treated by trainee doctors, who also handled and dissected cadavers. He believed that sticking one's hands into a dead body and then directly into a laboring woman was dangerous. So, Semmelweis issued a mandate that hands must be washed between the two activities. And it worked! Rates of infection dropped from one in ten to one in a hundred within the first few months. Unfortunately, the finding was rejected by much of the medical establishment of the time. One of the reasons it was so hard to get doctors to wash up? The stench of "hospital odor" on their hands was a mark of prestige. They called it "good old hospital stink." Quite simply, decayed corpse smell was a badge of honor they had no intention of removing.

What happens to soldiers who die far away in battle, or whose bodies are never found?

There are questions in this book that are more modern, e.g., "What happens if you die on a plane?" or "What happens to an astronaut body in space?" But other questions, like this one, have *been* questions for thousands of years.

Before the nineteenth century, long-distance transport of fallen soldiers rarely occurred—especially if there were hundreds or thousands of casualties. If you were a grunt—a foot soldier, a dude on the front lines who was stabbed with a spear, sword, or arrow—you would likely be left behind. If you were lucky, you might be given the dignity of a burial in a mass grave or a cremation, rather than being left to rot on the battlefield. The men who were brought all the way back home for burial tended to be the high-mucky-mucks: the generals, the kings, the famous warriors.

Take the British admiral Horatio Nelson. He was killed by a French sniper on the deck of his own ship during the Napoleonic Wars. His fleet won (congrats), but their leader was still dead, and required a hero's burial back home. So, to preserve him for the journey, his crew stuck Nelson's body in a barrel filled with brandy and aqua vitae (concentrated alcohol, literally "water of life"—ironic, no?). It took a month to sail back to Britain, and during the voyage Nelson's gases built up in the tiny barrel, causing the lid to pop off, terrifying the watchman.

Ever since, a rumor has persisted that the ship's sailors took turns sneaking sips of the alcoholic "embalming fluid" from Lord Nelson's barrel. Allegedly, they would use pieces of macaroni as tiny straws, and then top off the brandy barrel with less desirable wine to hide their crime. Personally, I would have stuck with drinking the wine that didn't have a corpse bobbing around in it, but British soldiers at the time were known for going to extremes in their quest for liquor.

For most of Western history, wars were fought by hired professional soldiers and men forced into battle. If they won, the credit for their victories would go to kings or, later, great generals. By the beginning of the twentieth century, Americans started to see bringing the bodies of ordinary soldiers home as the "humane" thing to do. President William McKinley even organized teams to bring back soldiers who died fighting Spain in Cuba and Puerto Rico.

That doesn't mean the procedure has worked-out-no-problem ever since. Far from it. After World War I, America was like, "Okay, France, we're coming over to excavate the mass graves containing all our dead soldiers, see you soon." France, working hard to rebuild, didn't want to be disturbed by these huge excavation projects. Many Americans who lost sons and husbands weren't so excited about the graves being disturbed, either. President Theodore Roosevelt himself wanted the remains of his son, a military pilot, to remain in Germany, and said, "We know that many good persons feel differently, but to us it is painful and harrowing long after death to move the poor body from which the soul has fled."

In the end, the U.S. government sent a survey to each family to see what they wanted done for their dead. As a result, 46,000 soldiers were returned to the United States, and 30,000 soldiers' bodies were buried in military cemeteries in Europe. To this day, there are heartwarming stories of Dutch and Belgian families adopting graves of American soldiers from both world wars, visiting and bringing flowers more than a century later. (Remember *that* when you don't want to go to the cemetery on Grandma's birthday.)

As your question suggests, however, it's not always an option to bring home a perfectly intact and identifiable soldier. There are still 73,000 missing bodies of World War II American service members. More than 7,000 service members are still missing from the Korean War, which ended in 1953. The majority of those bodies are likely in North Korea, where diplomatic negotiations are, shall we say, touchy at the moment.

Since 2016, the American agency in charge of tracking down and identifying missing and lost bodies and remains has been the Defense POW/MIA Accounting Agency. The researchers at the agency rely on eyewitness and historical accounts, forensics, and anything that can help them narrow down a geographic area where there might be remains. If they believe a certain location will contain remains, the agency dispatches a recovery team, which carries out the scientific research and retrieval. It sounds sort of glamorous (international body mysteries!), but much like working at a funeral home, the actual labor mostly involves obtaining permits and permissions, then working with the local government and families to make sure things go smoothly.

Let's talk now about what would happen if a soldier were to die tomorrow. How would the body be handled? I will use the United States military as an example. The U.S. (for better or worse) is a military superpower, meaning we don't have soldiers fighting and dying on our home soil. Rather, our soldiers often kill and are killed in distant lands. Even if you disagree with military policy, or with war in general, you can probably understand the desire of the dead soldier's family to see the body brought home, or at least decently buried or cremated.

Here's what happens now. Almost all the remains of American service members killed in the recent Iraq and Afghan conflicts have come through the Dover Port Mortuary, located at the Dover Air Force Base in Delaware. The mortuary is overseen by the Air Force and is the world's largest. Its facilities have the potential to handle one hundred bodies a day, and it has freezer storage for about a thousand more. This amazing capacity made it the first choice to receive dead bodies from the mass suicide at Jonestown, the bombing of Marine headquarters in Beirut, the *Challenger* and *Columbia* spaceship disasters, and the September 11 attack on the Pentagon.

When bodies first arrive at the Dover Port Mortuary, they are taken to the Explosive Ordnance Disposal Room, to make sure they're not holding concealed bombs. The bodies are then officially identified, using full-body X-rays, FBI fingerprint experts, and DNA tests to match blood samples given by service members before deployment.

The goal of the morticians is to make the soldiers' bodies viewable by their family. About 85 percent of families are able to hold a viewing. But with roadside bombs and other violent ways to die, there are cases where there's too little of the body left to reconstruct. Those remains are wrapped in gauze, sealed in plastic, and wrapped again in white sheets inside a green blanket. Finally, a full uniform is pinned on top. When families receive the incomplete bodies, they can choose whether they'd like additional remains (if any are found) to be sent to them in the future.

What happens when the body arrives at Dover Port, and when the body is returned to the family, is very ritualistic, very ordered, very . . . military. The mortuary has every uniform on hand for every possible branch and rank of soldier. That's every set of pants and jacket, but also every bar, stripe, flag, badge, cord, you name it. When the body is flown home, a soldier is assigned to fly with the body and salute as the body is loaded in and out of the plane (even if the body is just being transferred between flights). Then, there is the American flag, draped over the casket. There is a specific way of folding and draping the flag. Funeral directors' groups online have knock-down drag-out fights over what they say are improperly draped flags (correct way: blue field of stars over the person's left shoulder.)

When a body comes into my funeral home, I often know a great deal about the person: how they died, what they did for a living, even their mother's maiden name. That's because at a typical mortuary, the same funeral director may file the death certificate and prepare the body for a viewing. This is not the case at Dover Port Mortuary. The mortuary workers there are divided into two groups. One group handles the soldier's personal effects and identifying information, and the other group handles the physical bodies. The idea here is that no worker should become too personally familiar with any particular dead soldier. On one hand, that seems sad and depersonalized, but on the other, according to *Stars and Stripes* magazine, in 2010 "one in five mortuary affairs specialists sent to Afghanistan or Iraq returned with symptoms of post-traumatic stress disorder." That kind of bureaucracy and separation might be needed to deal with the trauma of war.

Can I be buried in the same grave as my hamster?

I get it—you love your hamster. Rightly so. Your hamster is probably more fun than most of the people you know. And a better conversationalist. People are awful, is what I'm saying.

You're not alone in wanting to give Hammibal Lecter a proper burial. People have wanted to give their animals a dignified send-off since—well—forever. In 1914, workers found a 14,000-year-old grave near Bonn, in Germany. In the grave were two humans (a man and a woman) and two dogs. One of the dogs was just a puppy, a very sick puppy infected with a canine virus. There is evidence that the humans cared for the puppy for some time before it finally died, which, given the virus, likely entailed keeping it warm and cleaning up its diarrhea and vomit. We don't know why the two dogs ended up buried with the humans. Maybe they were companions for the afterlife, symbolic somehow, or maybe the humans just really adored them. (Do *you* clean up diarrhea for things you don't adore?)

Everyone knows about the mummies of ancient Egyptians, but less is known about their exquisite animal mummies. Egyptians performed mummification on cats, dogs, birds, even crocodiles. Some animal mummies may have been offerings to gods or guardians, or even food for the afterlife, but cats were also

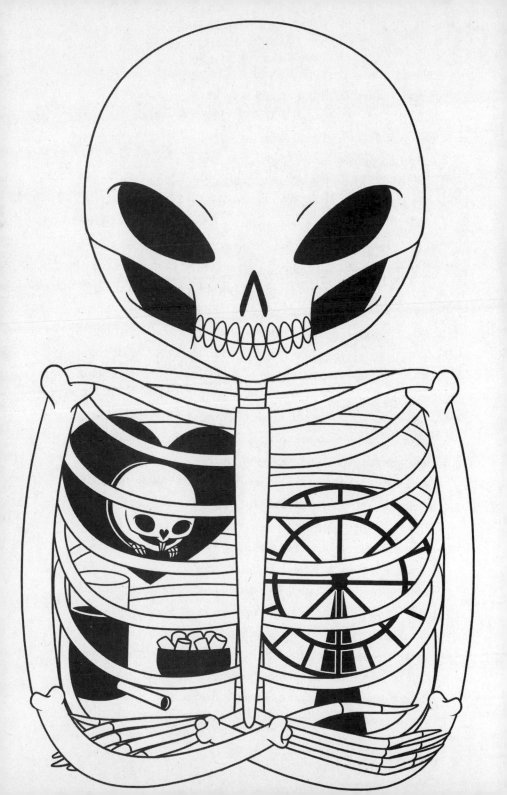

beloved domestic pets and were delivered to their owners' tombs (after a natural death) as companions in the great beyond.

In the late 1800s, over 200,000 of these (mostly cat) mummies were excavated from a massive burial site in central Egypt. A British professor wrote: "An Egyptian fellah from a neighbor-ing village . . . dug a hole, somewhere in the level floor of the desert, and struck—cats! Not one or two, here and there, but dozens, hundreds, hundreds of thousands, a layer of them, a stratum thicker than most coal seams, ten to twenty cats deep, mummy squeezed against mummy tight as herrings in a barrel." Cat mummies were wrapped, often painted and decorated with great care, and even given hollow bronze cases in which to spend eternity.

Nowadays, you're Crazy Cat-Lady Caitlin if you to want to be snuggled up next to Mr. Paws forever. But that's the wrong way to look at it! Humans have a long, rich history of being buried alongside their animals, and you and your hamster should be no different.

Say you died, and your family came to my funeral home to make arrangements for your burial. "He just loved Hammibal Lecter!" they say. "Can the hamster go in the casket?" My first question: is Hammibal Lecter also dead? If he's not, I'll need to do some thinking. I like to keep an open mind, but I'm not totally comfortable with the euthanasia of healthy animals for burial. All through human history, animals have been sacrificed to join their masters in the underworld, but that doesn't make it an ethical idea in the twenty-first century. Let's assume your hamster is already dead: taxidermied, just bones or ashes, or being kept in the freezer for this very occasion.

Technically, per the laws of the state of California, I am not allowed to slip Hammibal into your pocket, even if he's just a small pouch of cremated Hammibal remains. I'm not allowed to "bury" an animal in a human cemetery. Would I do it anyway? Umm, no comment. *(tiny paw extends from your suit pocket)*

Other U.S. states are more progressive on the issue of humans and animals being buried together. New York, Maryland, Nebraska, New Mexico, Pennsylvania, and Virginia are good examples. These states will allow your hamster (whole-body or cremated) and you, its owner, to be buried together. In England, "joint" human and animal cemeteries allow you to be buried *near* Hammibal, and in the last decade some joint cemeteries have even started allowing Hammibal to go directly into your grave.

The law in most states, even California, used to be more loosey-goosey on where animals could legally be buried. Take a stroll through some of America's oldest cemeteries and you'll see graves marking the burial sites of creatures like Moscow, the Civil War horse, who was interred at Sand Lake Union Cemetery in New York. Or dog actor Higgins, a.k.a. Benji I, buried at Forest Lawn Memorial Park in the Hollywood Hills.

You are not alone in wanting, nay *demanding*, to be buried with your beloved pet. There is a movement called "whole-family cemeteries," which argues that your whole family (mom, dad, hamsters, iguanas) should all be allowed to be buried in the same place. And it's catching on. Unfortunately, in many states the burial of pets in a human cemetery remains frustratingly illegal. Such laws assume that it's disrespectful to have animals in *human* cemeteries that should be reserved for *human* burial—that the presence of animal remains cheapens the human ritual of burial.

I understand this argument. There are religious and cultural reasons you might not want to be buried with someone else's beloved dog or family pig. Plus, with cemeteries running out of room in many major cities, humans are rightfully concerned that a primo corner plot might be taken by Cuddles the Great Dane.

I'm all for choice in death. If you want to be buried with no animals, you should have that. If you want to be buried with animals, you should have that. More places than you'd expect have animals-getting-buried-with-people on the legislative agenda. So yes, it's not out of the question that you and your furry friend could be buried together, running hand-in-paw on that great big hamster wheel in the sky. No matter what the local laws say, there may be a funeral director who's willing to sneak your pet's ashes into your casket.

Not me, of course. Next question.

Will my hair keep growing in my coffin after I'm buried?

TV host Johnny Carson once joked, "For three days after death hair and fingernails continue to grow, but phone calls taper off." Oh Johnny, you cad! You can rip my smartphone out of my cold dead rigor-mortised hands, thank you very much. We'll just see about those phone calls from the beyond.

But what about hair and fingernails growing in the grave? If we dug up your body thirty years after you died, will we find some desiccated skeleton topped with wild glam metal hair and six-and-a-half-foot-long fingernails?*

This image sounds way spooky and I wish I could tell you it is true. Alas, this is yet another death myth—a death myth that has been in pop culture from the very beginning. In the fourth century BC, Aristotle wrote that "hair actually goes on growing after death." He clarifies that for the hair to keep growing, the hair has to already exist, like a beard. If you're an old guy with a bald spot, no dice on filling that gap postmortem.

This myth has continued for over two thousand years. Into

---

* The nail length of the current world record holder.

the twentieth century, reputable medical journals were still reporting stories like "Thirteen-year-old girl dug up from the grave in Washington DC found with hair down to her feet" and "Doctor reports hair from inside a coffin busted through the seams and is shooting out the sides." The idea of hair vines winding their way through the dirt sounds cool, but it still didn't happen.

I'm not going to blame this myth solely on books and medical journals and movies. The myth persists because it looks like hair and nails are growing after death. When people observe something happening right before their eyes, it feels like Science 101. But what if what you are seeing isn't what you think you're seeing? Let me explain.

When you're alive, your fingernails grow about 0.1 mm every day. "Excellent, more for me to chew on!" my gross mind thinks. (Don't bite your nails, kids.) Hair grows a little less than 0.5 mm every day.

But you have to be alive for that hair and nail growth to happen. In order to grow hair and nails, your body needs to be making glucose, which in turn allows new cells to be created. In your fingernails, the new cells shove the old cells forward, growing the nail. It's almost like pushing the toothpaste out of the tube. It's the same story with your hair. New cells created at the base of the hair follicle push the old hair out of your face and head. But that whole glucose-making cell-creating process stops after you die. Death means no new nail, no luscious new locks.

So, if nothing is growing, why does it look like your hair and nails are getting longer? The answer has nothing to do with your luscious locks and everything to do with your skin, the largest organ of your body. Skin often becomes dehydrated after death. The once plump, living skin shrivels and retracts. If you have ever seen a time-lapse video of a ripe peach shriveling up over the course of a week, it's a lot like that.

When the skin on your hands dehydrates after death, the nail beds pull back, revealing more nail. The nails might seem longer, but it's not the nail growing, it's the skin revealing additional nail that was there all along. Same principle with hair. It might look like a dead man is growing out his stubble, but that's not real hair growth. It's his face drying out and shrinking to reveal the stubble. In short: it's not that there's more hair or nails, it's that there's less plump, living skin around the hair and nails. Two-thousand-year-old mystery solved.

Fun fact: to prevent the look of dehydrated hands and faces, sometimes funeral directors give the face a moisturizing facial and the nail beds little manicures for a viewing. The postmortem spa treatment we all deserve.

Can I use human bones from a cremation as jewelry?

When most people think of cremation, they picture a funeral director meeting the family and handing over an urn full of a gray, fluffy, sandlike substance. These ashes, or cremated remains, are now ready to sit in the back of a closet (sadly this happens more often than you'd think), or be scattered in the sea, or blown back in your face like in *The Big Lebowski*. These ashes were somehow once Dad, but what part of Dad are they, exactly? Well, kids, to be clear, ashes are dad's ground-up bones. *(metal riff plays)*

You probably sort of knew that already if you've made it this far in the book. But what you might not know is that ashes don't come out of the cremation machine looking like a bag of powdered sugar. During the intense heat of the cremation, all of the soft, fleshy, organic parts of Dad burn away, going up the chimney like a reverse Santa Claus. What the crematory operator pulls out of the machine is dad's inorganic bone. By that, I mean actual big chunks of bone: femurs, skull fragments, ribs.

Depending on what country you live in, one of two things happens to the bones after a cremation. The first thing that can happen is nothing. Those chunks of bones are given directly back to the family in a large urn. One of my favorite death rituals, the *kotsuage*, from Japan, involves the careful handling of cremated skeletons.

Japan has the highest cremation rate in the world. After a body is cremated there, the bones are allowed to cool before they are laid out for the family of the deceased. Starting at the feet and working up toward the head, the family uses long white chopsticks to pick pieces of bone out of the ash to deposit in an urn. They start from the feet and work their way up to the head because they don't want to make the dead person spend an eternity upside down.

Sometimes, larger bones, like thigh bones, require two people to pick up the bone at once. And sometimes the family will pass bone fragments to one another, chopsticks to chopsticks. This is the only time passing something between chopsticks is not considered rude. If you were to do this in public, with, say, a pork sparerib at a restaurant, it would be like bringing a funeral ritual to the dinner table. Total faux pas.

Compared to the elegance of the *kotsuage*, the second thing that might happen to a body after cremation seems more violent. In the Western world, chunks of bone are pulverized by a machine called the Cremulator. The bones go in a metal pot, whir around with sharp, fast-moving blades, and voilà, we've got ashes.

If you live in a country that typically requires the grinding of the bones, can you request that they be returned to you unground instead? United States funeral laws require that the crematory *must* pulverize the bones to an "unidentifiable" size. There seems to be a lot of concern about the deceased's family being able to identify a piece of Grandpa's hip. That said,

I know a few crematories that have given unpulverized bones back to families due to religious or cultural reasons. ("No Cremulator for Dad, thanks.") It never hurts to ask.

Let's address the elephant in the room: the jewelry. I'm assuming the bone jewelry is to honor Dad and not some dark revenge fantasy where you *destroy* him bone by blessed bone. Here's the problem: if you were trying to make jewelry out of your dad's post-cremation bones, destroying them is exactly what you'd likely end up doing.

In making bone, calcium phosphate and collagen bind together. The resulting bone is so strong that individual bones can successfully be used in jewelry (in fact, some people enjoy wearing an animal bone brooch). But those bones were cleaned by decomposition, the sun, dermestid beetles, etc. They weren't put through the cremation process.

Bones that are subjected to the 1,700-degree cremation chamber don't fare as well. That kind of heat not only causes tissue and smaller bones to disintegrate entirely, but it also undermines the strength and integrity of larger bones.

Whatever bones survive the process will be dehydrated. They lose volume and sustain permanent damage to their outer layers and internal microstructures. The hotter it gets inside the cremation chamber (the bigger the body, the hotter it can get), the more damage to the bones.

The bones we pull out post-cremation are cracked, brittle, and deformed. So brittle that the crematory operator could crush them to dust in their hand. Think: a very stale cookie. While the bones would be basically recognizable, they would be peeling, chipping at the edges, and would fall apart if you tried to, say, string them into a necklace.

If you're really determined to make jewelry out of your family member's remains, consider the ashes a viable candidate. There are *thousands* of cremated-remains jewelry options on the market. Little vials, glass pendants—send those ashes to a reputable dealer and in a few short weeks you could have yourself a cremated remains necklace, or ring, or almost any kind of jewelry. If you dream it, it can jewelry.

Sorry to disappoint on the human-bone jewelry front. But just feel lucky you don't live in Germany! Here's a story my friend Nora Menkin, a funeral director, told me. She had a family come to her for help when the father died on their German vacation. Retrieving the ashes proved to be a surprisingly long, complicated process that involved lots of Google translate (the German words for "urn" and "ballot box" are apparently very similar), mostly because Germany has very strict laws over who can and cannot handle cremated remains. Basically, *only funeral directors can.*

Not only are families barred from bringing Dad's cremated remains home, funeral directors are the only ones who have the authority to transfer ashes between urns, *and* the only ones who can bring the ashes to the cemetery for burial. Furthermore, the law requires that all ashes be buried. You can forget about jewelry, let alone your necklace made from Grandma's femur.

Obviously you, gentle reader, are not squeamish about cremated remains (and neither are the Japanese). If the bones are really important to you, look up the laws where you live, and don't be afraid to ask the funeral director or crematory. Just don't count on Dad's rib making a successful or attractive hair barrette.

Did mummies stink when they were wrapped?

The first mummies in Egypt were created by accident. Lower Egypt (where most of the pyramids are) doesn't get much rainfall. Combine that dryness with the sun and sand, and you have a recipe for natural mummification. It wasn't until around 2600 BC, over 4,600 years ago, that the ancient Egyptians decided to mummify their dead on purpose.

The most famous mummies—think, King Tutankhamen—are from roughly 3,300 years ago. This is the charismatic mummy we all know, a curled, leathery body wrapped in linen, kept for thousands of years in a golden sarcophagus in a fortress-like tomb, complete with a dreaded pharaoh's curse that befalls you if you dare disturb the grave.

I'm kidding about the curse, although, seriously, don't desecrate graves, kids.

Almost everyone who has ever lived on Earth (over 100 billion of us) has decayed or burned away into particles and atoms, lost to history. What's so thrilling about these mummies is that not only are they still around, but their bodies are so well-preserved that we can learn an extraordinary amount about how ancient Egyptians lived—everything from how they

died, to what they looked like, to what they ate. A mummy is a time capsule from an ancient culture.

Okay, enough geeking out about mummies. Let's get to the point: were they stinky as they were being wrapped up? The answer is yes, they were stinky after they died. But by the time

they were being wrapped in hundreds of yards of linen, not as much. The process of ancient embalming wasn't quick. It didn't go: King Tut dies, wrap him up, put him in the tomb, call it a day. The process of mummification could last months.

The first step was the removal of the body's internal organs. This is where things likely got smelly. In my job, I've had to remove organs from dead bodies to repair a corpse after an autopsy. If the person has been dead for a week or more, with organs decomposing and gas building up inside, opening the stomach cavity can be an unpleasant experience. You are met with a wall of sweet, rotting stink. I imagine it would have been similar when, a few days after death, ancient embalmers removed the liver, stomach, and lungs and put them in special receptacles called canopic jars (jars topped with animal and human heads), which would be buried with the body later.

You've probably heard that the other major organ removed during mummification was the brain. And sometimes it was. Ancient undertakers used a hooked tool that either went up through the nose or through a small hole in the base of the skull. In 2008, a CT scan of the head of a 2,400-year-old female mummy found a brain removal tool still stuck in the back of her skull. (I certainly hope that embalmer got a negative Yelp review.) But other mummies have been found with the brain still intact inside the skull. Brain removal through the nose would have been a difficult process, and it wasn't available to everyone.

In the next step, the newly eviscerated body was dried out. The future mummy (now without its organs) was packed inside and out with natron, a salt mixture that the Egyptians collected from dry lake beds. The sodium carbonate and sodium bicarbonate in natron would absorb water and desiccate the body over the course of thirty to seventy days. All the enzymes that work to dissolve our dead flesh require water, so dehydrating the body like beef jerky prevents those enzymes from doing their sinister decay work.

Now, a regular dead body, untreated and untouched, would smell truly vile if left out in a hot climate like Egypt's over a period of seventy days. After the embalmer removed the organs and packed the body in salt, I imagine the body didn't smell great, but it wouldn't smell anywhere close to as bad as a naturally decaying body.

After the natron was removed, the embalmers filled the body cavities with sawdust, linen, and pleasant-smelling substances like cinnamon and frankincense. It's possible that at some moments the dry body may have even smelled . . . sorta nice? Like a Christmas candle. Or a pumpkin-spiced mummy.

Now the mummy was ready for wrapping. This part of the process involved meticulous application of various oils and resins from coniferous plants (which may also have helped with smell). Round and round the linen went, around body's fingers and toes individually and then binding the whole feet and hands. Keep in mind that this embalming had a religious purpose. It was believed that the soul had multiple parts, and those parts resided in different regions of the body. If the body couldn't be preserved, where would the soul find a home? But the people who had these elaborate corpse treatments and prayers and tombs created for them were mostly those who could afford it. (Cough, rich people.)

So the answer to your question is, by the time the mummy wrapping was happening, often more than a month into the process, the corpse would have been disemboweled and dried and stuffed to the point that the smell probably wouldn't have been awful. Unbothered by the scent, Egyptians would then proceed to the next step: placing the body into a sarcophagus for thousands of years. Now, you asked if the mummy smelled as it was being wrapped. But what about when it is *unwrapped* for study in the twenty-first century? Can a mummy carry a stench across the ages?

The good news is that, these days, mummies are unwrapped far less than they used to be. In the nineteenth century, Europeans were obsessed with Egypt. People in England would hold mummy-unwrapping parties, for which hucksters would sell tickets to the public, so they could watch ancient mummies being unwrapped (destroying the mummy in the process). So many Egyptian tombs were looted that mummies were ground up and used as brown paint for artists or added to medicines: "Take two mummy pills and call me in the morning."

Nowadays, scientists can learn just as much, if not more, from studying the mummies with technology like CT scans than they can from direct observation and dissection. Information can thus be gathered without damaging the very fragile 3,000-year-old mummy. As for the smell of an unwrapped mummy? It's been compared to old books, leather, and dried cheese. Which doesn't sound so bad either. Don't blame our ancient friends for the stench; it's those fresh, week-old decomposing corpses you have to watch out for.

At my grandma's wake, she was wrapped in plastic under her blouse. Why would they do that?

I think Grandma got leaky. It's not Grandma's fault, I'm sure she was a very neat, tidy person while she was alive. But the human body is full of fluids, fluids that become difficult to control after death. The funeral industry has a term for it: leakage.

Funeral directors hate leakage. Leakage is their nightmare. Funeral directors do everything in their power to stop fluids from making a surprise appearance. But after a death occurs, some bodies are just leakier than others. Say your family wanted an expensive wake. Everyone from Grandma's church and family from three countries have flown in to see her body at the viewing. Grandma is embalmed and lying in a casket with a pale violet crepe interior, wearing her favorite peach-colored silk dress. In this situation, Grandma springing a leak is not an option.

So what are some of the ways funeral directors try to prevent leakage? First, you have to determine the source of the fluid. The most obvious places Grandma will leak are, not to be too crass here, her preexisting holes. Her mouth, her nose, her vagina, and her rectum. Usually the first things to leak are liquids and other sticky things that the body was designed to excrete: urine, feces, saliva, phlegm, the delightful list goes on. If the funeral director is worried about a poopy surprise (the least fun of all surprises), then diapers and absorbent pads will be placed around Grandma's lower regions. Decomposition in Grandma's stomach might produce a substance called "purge," an unattractive liquid like coffee grounds that sometimes exits via the nose and mouth. Before the viewing, the funeral director may suck out Grandma's mouth and nostril cavities with a small aspirator (sucking machine), and pack the nose and mouth with cotton or gauze to catch anything trying to escape.

Those are typical leakage problems, but you're wondering why Grandma was actually wrapped in plastic under her clothes. There are several reasons the funeral director may have gone that route. And no, it wasn't to keep her fresh, like a shrink-wrapped grocery store vegetable. Did Grandma die after an extended stay in the hospital or after a long illness? If the answer is yes, then when she was brought to the funeral home she may have had open wounds on her arms and legs, anything from surgery incisions to needle holes from her IVs, or the everyday chronic wounds the afflict the elderly due to illness or aging skin. Cuts or wounds that would heal easily on young skin like yours have a much harder time healing in someone very sick or much older. And keep in mind that, after death, a wound doesn't scab or start to heal. Wounds you have when you die stay fresh wounds. So perhaps the funeral director used gels or powders to dry the wounds and then placed plastic wrap around them to stop them leaking.

There are also multiple medical conditions that could cause Grandma to leak. If she had diabetes or was overweight, her circulation, especially down to her legs, might not have been great.

This bad circulation can cause water blisters or skin problems. Even worse (for the funeral director) would be if Grandma had edema. Edema is not a word we hear often, but it strikes fear into the heart of a funeral director. It refers to abnormal swelling in the body, when fluid collects under the skin. There are many reasons edema develops. Maybe Grandma had cancer and was getting chemotherapy or other medications; maybe her liver or kidneys were failing; maybe she had an infection. Whatever the reason for the edema, the funeral director would have to be very careful in handling her paper-thin, swollen, weepy skin. In fact, edema would have caused a 10 percent increase of fluid in her body (we're talking gallons of liquid here). That's a lot of extra fluid to keep contained.

Sometimes funeral directors, worried about fluids leaking out of the skin, will dress the body in head-to-toe clear vinyl garments called unionalls, which look like those adult onesies. Funeral homes can also purchase the parts individually—a vinyl jacket, plastic capri pants, or synthetic booties, if only one area of the body is leaking. The funeral director then puts the person's clothes on over the vinyl garments. It's interesting to see how different mortuary supply companies advertise their corpse onesies: "Will not crack, peel, or deteriorate!" "Second to none in the industry!"

Maybe what you saw was one of these unionalls. But many funeral directors turn to good ol'-fashioned Saran Wrap, the clear plastic wrap you use to cover your leftovers. If it ain't broke, don't fix it. Some of the very nervous (or careful) in my profession use shrink wrap, heating it up with a hairdryer to seal it, then put the unionall garment over the top of the shrink wrap.

Something to think about (and something I and my staff think about a lot) is why we are so afraid of a dead body leaking a little bit? We want to control our corpses, but just as you can't stop a newborn baby from crying, you can't stop a dead body from doing what a dead body does. Our funeral home takes a more natural approach to preparing a dead body, which means we don't use chemicals to preserve it, and we don't use chemical powders on the body. If we are doing a natural burial for the family, we wouldn't be allowed to do those things even if we wanted to. The body has to go into the ground wearing nothing but unbleached cotton clothes.

So, if your grandmother had come to our funeral home, we wouldn't have wrapped her in Saran Wrap. But we would be forced to have a hard, honest conversation with you about what you were going to see when you viewed her body—whether it was Grandma's wounds or weeping skin. Keep in mind, some of these plastics and wraps were adopted by funeral homes over the years because of lawsuits. Families sued because the eggshell interior of the (very expensive) casket or that peach-colored silk dress became soiled or ruined because the funeral director didn't do their job to "protect" the body.

Morticians aren't magicians, and a dead body will never behave 100 percent, no matter how much Saran Wrap you use. There are all different kinds of funeral homes and different philosophies about what makes a "good" corpse. For me, it's a natural corpse. But if your family had all the church folks and all the relatives there for the wake, they may have wanted to keep everything under control and wrap Grandma up. That's for the family to decide.

Acknowledgments

This book would not exist without hundreds of questions from morbid little angels. I'm thankful for your curious minds and understanding parents.

My editor, Tom Mayer, usually publishes highbrow fare (think, Afghanistan or the history of jazz), but, for me, will go four rounds of edits on corpse poop. My gratitude is obvious.

My agent, Anna Sproul-Latimer, has three flawless children who are almost old enough for Auntie Caitlin to repay Anna's years of hard work by teaching them the stages of putrefaction.

I am grateful to have tricked a whole team of professional book publishers at W. W. Norton into taking very seriously questions like, "Should the title be *Grandma's Viking Funeral* or *Cat Ate My Eyeballs?*" Thanks to my immediate team: Erin Lovett, Steve Colca, and Nneoma Amadi-obi. And my big team: Ingsu Liu and Steve Attardo, Brendan Curry and Steven Pace, Elisabeth Kerr and Nicola DeRobertis-Theye, Lauren Abbate and Becky Homiski, and Allegra Huston.

I would have been a wraith wandering the moors of confusion without the keen eyes and research skills of Louise Hung and Leigh Cowart.

The people I get to call both experts and friends, like Tanya Marsh, Nora Menkin, Judy Melinek, Jeff Jorgenson, Monica Torres, Marianne Hamel, and Amber Carvaly.

The whole Order of the Good Death team, especially Sarah Chavez for protecting me from the dark, cruel world.

Dianné Ruz for being a diabolical genius.

And finally to Ryan Saylor, the shroud to my casket.

Rapid-Fire Death Questions!

When choosing questions for *Will My Cat Eat My Eyeballs?* I was faced with an embarrassment of riches. I had hundreds of fantastic questions to select from. Some questions didn't make the cut simply because, cool as they were, they didn't require a robust, several-page answer. (My publisher insisted the chapters be longer than a paragraph.)

To finally give those their due, I present RAPID-FIRE DEATH QUESTIONS.

Is it bad for the environment to be buried in a large dragon costume?

Depends what the dragon costume is made out of! "Green" or "natural" burial grounds, for the most part, only allow bodies to be buried wearing natural fibers, such as unbleached cotton. So that bodacious disco polyester sparkle bodysuit? Sorry, it's out. Same goes for most of the polyester and velour dragon costumes I found for sale online. (A very adorable search!) But perhaps with your costuming skills you've created a dragon costume using natural materials. Look at you go, you mythical hemp-dragon corpse creature, you. Although if you're that interested in fire-breathing, you might consider cremation.

Can honey stop your body from rotting?

Yes! Honey itself usually doesn't rot, so it is a perfect substance to preserve a body for the long haul. The intense level of sugar in honey stops bacteria from eating the body, and because moisture can spoil honey, it contains an enzyme that combines glucose and excess water to create hydrogen peroxide, which makes honey antiseptic. Move aside formaldehyde, there's a new embalming kid in town! People have preserved things, including bodies, in honey throughout human history, everywhere from ancient Egypt to modern-day Myanmar. Alexander the Great was said to have been embalmed in honey, though the location of his grave is one of the great mysteries of archaeology. So, honey works. But for some reason, being plopped in a vat of honey hasn't become as popular as other entombing methods. The honey lobby should get on this.

What happens if a cremation machine breaks during cremation?

I don't know and I hope I never have to find out.

What is the most unique insect that feeds on a body while it's decomposing?

Dermestid beetles get most of the attention, but corpses are also visited by other beetle families, including dung beetles, clown beetles, and carrion beetles. Lately I've been interested in spider beetles (*Ptinus*), which show up at the very end of the decomposition process, when the corpse has been taken down to bones. Timeline-wise, spider beetles might not arrive until years after the death has occurred. By this point, the only thing left is the waste that previous scavenging insects have left behind (poop, pupal insect cases, more poop). That excrement, spread thinly over the bones, is what the spider beetles are into. "Sign me up for that bone poop buffet," they say. There's truly something for everyone, folks.

If you're left in the sand in the desert, does the sun shrivel you up?

If a corpse is unburied, just lying on the sand in the desert, it's going to dry up, or desiccate, quickly. The sand acts as a desiccant, wicking away moisture, kind of like cat litter or rice. (You know how if you drop your phone in the toilet they tell you to nestle it in rice overnight to dry it out? In this scenario, you are the phone.) Any clothes the corpse might be wearing will also pull moisture away from the body, speeding up the drying process. Bugs and flies go bonkers trying to eat up that putrefying flesh and tissue while things are still moist, because after a while, tissues are going to get so dried out and hard that even bugs won't be able to eat through them. Eventually all that's left of a body are bones and brittle skin, similar in texture to parchment paper. At this point, the remains are mummified— and will have turned bright orange or red (instead of a corpse's usual grayish brown). If left untouched, such a desert mummy could theoretically last for years.

An Expert Answers: Is My Child Normal?

Being an expert on corpses doesn't make you an expert on children's psychological fears and anxieties around death. Before publishing this book, I worried the medical community would go, "Hey, wait a minute, why is this random mortician talking to children about death? She'll infect them with terror!"

Fortunately, they haven't. At least not yet. The medical consensus is that honest and specific conversations with children about death can actually help with their death fears. I asked my friend Dr. Alicia Jorgenson, a child and adolescent psychiatrist in Seattle, to read a draft of the book to make sure I wasn't poisoning the children with my maggot enthusiasm.

But for those parents out there still wondering, "Is my morbid Marky . . . normal?!" I present to you my conversation with Alicia.

CD: Do you often have children in your office who express fear or worry about dying?

AJ: It is not very often that a kid will tell me outright, "I am scared about dying." It is more likely they will talk about worries or fears about their health, or their parent's health, or germs and contamination.

So, the fear of death, especially in younger children, may show up as health concerns?

Exactly—health concerns are a very common manifestation of death anxiety. Interestingly, some kids may not even verbally express worries about their health. Instead, they can present with stomachaches or headaches as the first symptoms of an anxiety disorder. I have also seen kids develop worries about falling asleep—especially if they have heard that someone "passed away in their sleep."

What other, seemingly more common fears might actually be connected to fear of death?

Younger kids do not understand more euphemistic terms for death like "passing away" or "losing someone." Using this language can be confusing when they hear these words in other contexts—like a sibling getting "lost" in the grocery store. Or even that a person "died in a hospital." This could lead to a fear that if they go to the hospital, it means they will die. From a developmental perspective, younger children (aged between three and five years old) typically do not understand the abstract concept of death and instead see it as temporary or reversible—like in cartoon shows. Even slightly older kids are not yet the best at logical reasoning and tend to use a process of association to understand the world. Most experts think that the concept of death being final and irreversible is not understood until age nine (give or take a year). So, it is more helpful for parents and other adults to be careful in their language choice and to use the word "death" and spell out in concrete terms what that means.

What type of concrete language can be helpful?

Be honest and direct with simple language. I recommend using words like "death," "dead," and "dying" and being clear about what they mean. When someone dies, their body stops working, they stop moving, and they cannot feel anything. A dead person cannot come back to life. This may be a tricky concept for younger kids, but I do think you can honestly say that "Even though Grandpa has died, our memories of Grandpa can live on in our minds."

With children, is some anxiety around death normal?

Of course! Anxiety is a normal emotion that we all experience in stressful situations or with the unknown. It is going to naturally come up for kids when a death happens. Parents may be anxious about how they will explain death to their kids—also normal. It's good to be prepared for what you will say. It is also important to note that kids are going to look to their parents for how to model their own thoughts and behaviors around death.

Is there a time when preoccupation with death can become too much for a child?

For sure. Anxiety *disorders* are not normal anxiety. They happen when kids worry too much about something and change their behaviors to avoid doing things that make them anxious, which then interferes with their ability to function (for example, not wanting to go to school or not wanting to leave their parent's sight). By definition, an anxiety disorder involves an *unrealistic* fear of something bad happening. For example, this could be persistent daily worries about their parent dying even though they are not sick. Sometimes an anxiety disorder is initially triggered by something bad happening in the environment (like a death). But sometimes anxiety happens out of the blue. Often anxious kids have anxious parents—so there is both a genetic as well as an environmental predisposition to anxiety. The good news is that we have very good treatments for anxiety disorders in children and adolescents, typically starting with talk therapy and sometimes adding medications.

I worried all the time as a kid that my parent would die!

That's a lot of thinking about death, Caitlin! In this situation, it may be helpful to say something along the lines of "No one can promise that they won't die. But we can practice taking good care of ourselves by being healthy. So I expect that we will be together for a very long time."

Bottom line—it is appropriate to have anxiety or sadness when someone you love is sick, dying, or has died.

If you're an adult, is it ok to show sadness and grief to a child?

Adults all grieve in their own ways. But I do think showing sadness can be helpful. It can be confusing for kids to get the emotional or nonverbal message that "everything is fine" when it is not. It's okay to cry in front of kids—and it helps if you explain why you are sad. It is also okay to say "I don't know."

Do children experience grief the same way adults do?

Not exactly. Kids may not be able to talk about their grief in the same way that adults would. In general, I think about grief as a normal, albeit complicated, emotion that can happen after any kind of loss. For example, some form of grief may first be experienced with the loss of a favorite stuffed animal or a move to a new home. The death of a pet is a very common first exposure to mortality. In general, the closer a kid is to the person or animal who has died, the more intense the grief tends to be. For kids, like adults, there is not a "right" or "wrong" way to grieve.

What emotions or behaviors can you expect to see from children after a death?

Be prepared to be open to their range of emotions—including tantrums, sadness, and anxiety. It is also very normal for kids to just seem to move on. This can be confusing to parents. I recommend parents checking in with kids about their emotions but trying not to project their own emotions or grief onto the child. I often tell families who are grieving that keeping up with routines can be very reassuring for kids—for example, waking up at the same time, eating usual meals, playing, and going to school. Rituals around death can also be very helpful for kids (like adults). If they are going to be involved in a funeral, I do think parents and adults need to prepare a kid for what will happen, saying things like, "Grandma is going to look different in death than in life." I would not force a kid to go to a funeral, though, if they clearly do not want to. Sharing memories can also be very helpful, and asking kids about their own memories of the person who has died.

Sources

All websites accessed February 3, 2019.

When I die, will my cat eat my eyeballs?

Raasch, Chuck. "Cats kill up to 3.7B birds annually." *USA Today*, updated January 30, 2013. https://www.usatoday.com/story/news/nation/2013/01/29/cats-wild-birds-mammals-study/1873871/.

Umer, Natasha, and Will Varner. "Horrifying Stories Of Animals Eating Their Owners." *Buzzfeed*, January 8, 2015. https://www.buzzfeed.com/natashaumer/cats-eat-your-face-after-you-die?utm_term=.clnqjk9DM#.deQmAwq6K.

Livesey, Jon. " 'Survivalist' chihuahua ate owner to stay alive after spending days with dead body before it was found." *Mirror*, October 30, 2017. https://www.mirror.co.uk/news/world-news/survivalist-chihuahua-ate-owner-stay-11434424.

Ropohi, D., R. Scheithauer, and S. Pollak. "Postmortem injuries inflicted by domestic golden hamster: morphological aspects and evidence by DNA typing." *Forensic Science International*, March 31, 1995. https://www.ncbi.nlm.nih.gov/pubmed/7750871.

Steadman, D. W., and H. Worne. "Canine scavenging of human remains in an indoor setting." *Forensic Science International*, November 15, 2007. https://www.ncbi.nlm.nih.gov/pubmed/?term=Canine+scavenging+of+human+remains+in+an+indoor+setting.

Hernández-Carrasco, Mónica, Julián M. A. Pisani, Fabiana Scarso-Giaconi, and Gabriel M. Fonseca. "Indoor postmortem mutilation by dogs: Confusion, contradictions, and needs from the perspective of the forensic veterinarian medicine." *Journal of Veterinary Behavior* 15 (September–October 2016): 56–60. https://www.sciencedirect.com/science/article/pii/S1558787816301447.

What would happen to an astronaut body in space?

Stirone, Sharon. "What happens to your body when you die in space?" *Popular Science*, January 20, 2017. https://www.popsci.com/what-happens-to-your-body-when-you-die-in-space.

Order of the Good Death. "The final frontier . . . for your dead body." http://www.orderofthegooddeath.com/the-final-frontier-for-you-dead-body.

Herkewitz, William. "Could a Corpse Seed Life on Another Planet?" *Discover*, October 25, 2016. http://blogs.discovermagazine.com/crux/2016/10/25/could-a-corpse-seed-life-on-another-planet/#.WoNe0raZPEb.

crazypulsar. "Vacuum & Hypoxia: What Happens If You Are Exposed to the Vacuum of Space?" *Indivisible System*, November 7, 2012. https://indivisiblesystem.wordpress.com/2012/11/07/what-happens-if-you-are-exposed-to-the-vacuum-of-space/.

Czarnik, Tamarack R. "Ebullism at 1 Million Feet: Surviving Rapid/Explosive Decompression." Available at http://www.geoffreylandis.com.

Can I keep my parents' skulls after they die?

Zigarovich, Jolene. "Preserved Remains: Embalming Practices in Eighteenth-Century England." *Eighteenth-Century Life* 33, no. 3 (October 1, 2009). https://doi.org/10.1215/00982601-2009-004.

Carney, Scott. "Inside India's Underground Trade in Human Remains." *Wired*, November 27, 2007. https://www.wired.com/2007/11/ff-bones/.

Halling, Christine L., and Ryan M. Seidemann. "They Sell Skulls Online?!: A Review of Internet Sales of Human Skulls on eBay and the Laws in Place to Restrict Sales." *Journal of Forensic Sciences* 61, no. 5 (September 1, 2016). https://www.ncbi.nlm.nih.gov/pubmed/27373546.

McAllister, Jamie. "4 Things to Do With Your Skeleton After You Die." *Health Journal*, October 5, 2016. http://www.thehealthjournals.com/4-things-skeleton-die/.

Inglis-Arkell, Esther. "So you want to hang your skeleton in public? Here's how." *io9*, June 6, 2012. https://io9.gizmodo.com/5916310/so-you-want-to-donate-your-skeleton-to-a-friend.

"Can bones be willed to a family member after death?" Law Stack Exchange, edited December 26, 2016. https://law.stackexchange.com/questions/16007/can-bones-be-willed-to-a-family-member-after-death.

Hugo, Kristin. "Human Skulls Are Being Sold Online, But Is It Legal?" *National Geographic*, August 23, 2016. https://news.nationalgeographic.com/2016/08/human-skulls-sale-legal-ebay-forensics-science/.

OddArticulations. "Is owning a human skull legal?" January 6, 2018. http://www.oddarticulations.com/is-owning-a-human-skull-legal/.

The Bone Room. "Real Human Skulls." https://www.boneroom.com/store/c45/
Human_Skulls.html.

Evans, Murray. "It's a gruesome job to clean skulls, but somebody has to do it."
October 30, 2006. https://www.seattlepi.com/business/article/It-s-a-gruesome-job-to
-clean-skulls-but-somebody-1218504.php.

Marsh, Tanya. "Internet Sales of Human Remains Persist Despite Questionable
Legality." *Death Care Studies*, August 16, 2016. https://www.deathcarestudies
.com/2016/08/internet-sales-of-human-remains-persist-despite-questionable
-legality/.

"Sale of Organs and Related Statutes." https://www.state.gov/documents/
organization/135994.pdf.

Vergano, Dan. "eBay Just Nixxed Its Human Skull Market." *Buzzfeed*, July 12,
2016. https://www.buzzfeednews.com/article/danvergano/skull-sales.

Shiffman, John, and Brian Grow. "Body donation: Frequently asked questions."
Reuters, October 24, 2017. https://www.reuters.com/investigates/special-report/usa
-bodies-qanda/.

Lovejoy, Bess. "Julia Pastrana: A 'Monster to the Whole World.'" *Public Domain
Review*, November 26, 2014. https://publicdomainreview.org/2014/11/26/julia
-pastrana-a-monster-to-the-whole-world/.

Will my body sit up or speak on its own after I die?

Berezow, Alex. "Which Bacteria Decompose Your Dead, Bloated Body?" *Forbes*,
November 5, 2013. https://www.forbes.com/sites/alexberezow/2013/11/05/which
-bacteria-decompose-your-dead-bloated-body/#637b6f3295a8.

Howe, Teo Aik. "Post-Mortem Spasms." *WebNotes in Emergency Medicine*, December
25, 2008. http://emergencywebnotes.blogspot.com/2008/12/post-mortem-spasms.html.

Costandi, Moheb. "What happens to our bodies after we die?" *BBC Future*, May 8,
2015. http://www.bbc.com/future/story/20150508-what-happens-after-we-die.

Bondeson, Jan. *Buried Alive: The Terrifying History of Our Most Primal Fear*. New
York: W. W. Norton, 2001.

Gould, Francesca. *Why Fish Fart: And Other Useless or Gross Information About the
World*. New York: Jeremy P. Tarcher/Penguin, 2009.

We buried my dog in the backyard, what would happen if we dug him up now?

O'Brien, Connor. "Pet exhumations a growing business as more people move
house and take their loved animals with them." *Courier-Mail*, May 4, 2014. https://
www.couriermail.com.au/business/pet-exhumations-a-growing-business-as-more

-people-move-house-and-take-their-loved-animals-with-them/news-story/58069b3e
d49b6c49f1a3f9c7c1d11514.

Ask MetaFilter. "How to go about moving a pet's grave." May 3, 2012. https://ask
.metafilter.com/214497/How-to-go-about-moving-a-pets-grave.

Berger, Michele. "From Flesh to Bone: The Role of Weather in Body Decomposition." Weather Channel, October 31, 2013. https://weather.com/science/news/flesh
-bone-what-role-weather-plays-body-decomposition-20131031.

Emery, Kate Meyers. "Taphonomy: What Happens to Bones After Burial?" *Bones
Don't Lie* (blog), April 5, 2013. https://bonesdontlie.wordpress.com/2013/04/05/
taphonomy-what-happens-to-bones-after-death/.

Can I preserve my dead body in amber like a prehistoric insect?

Udurawane, Vasika. "Trapped in time: The top 10 amber fossils." *Earth Archives*,
"almost three years ago" (from February 13, 2019). http://www.eartharchives.org/
articles/trapped-in-time-the-top-10-amber-fossils/.

Daley, Jason. "This 100-Million-Year-Old Insect Trapped in Amber Defines New
Order." *Smithsonian*, January 31, 2017. https://www.smithsonianmag.com/smart
-news/new-order-insect-found-trapped-ancient-amber-180961968/.

Why do we turn colors when we die?

Geberth,Vernon J. "Estimating Time of Death." *Law and Order* 55, no. 3 (March
2007).

Presnell, S. Erin. "Postmortem Changes." *Medscape*, updated October 13, 2015.
https://emedicine.medscape.com/article/1680032-overview.

Australian Museum. "Stages of Decomposition." November 12, 2018. https://
australianmuseum.net.au/stages-of-decomposition.

Claridge, Jack. "The Rate of Decay in a Corpse." *Explore Forensics*, updated January 18, 2017. http://www.exploreforensics.co.uk/the-rate-of-decay-in-a-corpse.html.

How does a whole adult fit in a tiny box after cremation?

Cremation Solutions. "All About Cremation Ashes." https://www
.cremationsolutions.com/information/scattering-ashes/all-about-cremation-ashes.

Warren, M. W., and W. R. Maples. "The anthropometry of contemporary commer-
cial cremation." *Journal of Forensic Science* 42, no. 3 (1997): 417–23. https://www
.ncbi.nlm.nih.gov/pubmed/9144931.

Do conjoined twins always die at the same time?

Geroulanos, S., F. Jaggi, J. Wydler, M. Lachat, and M. Cakmakci. [Thoracopagus symmetricus. On the separation of Siamese twins in the 10th century A. D. by Byzantine physicians]. Article in German. *Gesnerus* 50, pt. 3–4 (1993): 179–200. https://www.ncbi.nlm.nih.gov/pubmed/8307391.

Bondeson, Jan. "The Biddenden Maids: a curious chapter in the history of conjoined twins." *Journal of the Royal Society of Medicine* 85, no. 4 (April 1992): 217–21. https://www.ncbi.nlm.nih.gov/pubmed/1433064.

Associated Press. "Twin Who Survived Separation Surgery Dies." *New York Times*, June 10, 1994. https://www.nytimes.com/1994/06/10/us/twin-who-survived-separation-surgery-dies.html.

Davis, Joshua. "Till Death Do Us Part." *Wired*, October 1, 2003. https://www.wired.com/2003/10/twins/.

Quigley, Christine. *Conjoined Twins: An Historical, Biological and Ethical Issues Encyclopedia.* Jefferson, NC: McFarland, 2012.

Smith, Rory, and Anna Cardovillis. "Tanzanian conjoined twins die at age 21." CNN, June 4, 2018. https://www.cnn.com/2018/06/04/health/tanzanian-conjoined-twins-death-intl/index.html.

If I died making a stupid face, would it be stuck like that forever?

D'Souza, Deepak H., S. Harish, M. Rajesh, and J. Kiran. "Rigor mortis in an unusual position: Forensic considerations." *International Journal of Applied and Basic Medical Research* 1, no. 2 (July–December 2011): 120–22. https://www.ncbi.nlm.nih.gov/pmc/articles/PMC3657962/.

Rao, Dinesh. "Muscular Changes." *Forensic Pathology.* http://www.forensicpathologyonline.com/e-book/post-mortem-changes/muscular-changes.

Senthilkumaran, Subramanian, Ritesh G. Menezes, Savita Lasrado, and Ponniah Thirumalaikolundusubramanian. "Instantaneous rigor or something else?" *American Journal of Emergency Medicine* 31, no. 2 (February 2013): 407. https://www.ajemjournal.com/article/S0735-6757(12)00411-1/abstract.

Fierro, Marcella F. "Cadaveric spasm." *Forensic Science, Medicine, and Pathology* 9, no. 2 (April 10, 2013). https://www.deepdyve.com/lp/springer-journals/cadaveric-spasm-aFQAGR1PmQ?articleList=%2Fsearch%3Fquery%3Dcadaveric%2Bspasm.

Can we give Grandma a Viking funeral?

Dobat, Andres Siegfried. "Viking stranger-kings: the foreign as a source of power in Viking Age Scandinavia, or, why there was a peacock in the Gokstad ship burial?" *Early Medieval Europe* 23, no. 2 (May 1, 2015). https://www.deepdyve

.com/lp/wiley/v-iking-stranger-kings-the-foreign-as-a-source-of-power-in-v-iking-a SDfkk3w00D?articleList=%2Fsearch%3Fquery%3Dviking%2Bfuneral.

Devlin, Joanne. Review of *The Archaeology of Cremation: Burned Human Remains in Funerary Studies*, edited by Tim Thompson. *American Journal of Physical Anthropology* 162, no. 3 (March 1, 2017). https://www.deepdyve.com/lp/ wiley/the-archaeology-of-cremation-burned-human-remains-in-funerary-studies 0JPA0fEoP9?articleList=%2Fsearch%3Fquery%3Dcremation%2Bscandinavia.

ThorNews. "A Viking Burial Described by Arab Writer Ahmad ibn Fadlan." May 12, 2012.

https://thornews.com/2012/05/12/a-viking-burial-described-by-arab-writer-ahmad -ibn-fadlan/.

Spatacean, Cristina. *Women in the Viking Age: Death, Life After and Burial Customs*. Oslo: University of Oslo, 2006.

Montgomery, James E. "Ibn Fadlan and the Rusiyyah." *Journal of Arabic and Islamic Studies* 3 (2000). https://www.lancaster.ac.uk/jais/volume/volume3.htm.

Why don't animals dig up all the graves?

Hoffner, Ann. "Why does grave depth matter for green burial?" Green Burial Naturally, March 2, 2017. https://www.greenburialnaturally.org/blog/2017/2/27/why -does-grave-depth-matter-for-green-burial.

Harding, Luke. "Russian bears treat graveyards as 'giant refrigerators.'" *Guardian*, October 26, 2010. https://www.theguardian.com/world/2010/oct/26/russia-bears-eat -corpses-graveyards.

A Grave Interest (blog). April 6, 2012. http://agraveinterest.blogspot.com/2012/04/ leaving-stones-on-graves.html.

Mascareñas, Isabel. "Ellenton funeral home accused of digging shallow graves." *10 News*, WSTP, updated November 1, 2017. http://www.wtsp.com/article/news/ local/manateecounty/ellenton-funeral-home-accused-of-digging-shallow-graves/67 -487335913.

Paluska, Michael. "Cemetery mystery: Animals trying to dig up fresh bodies?" *ABC Action News*, WFTS Tampa Bay, updated October 30, 2017. https://www .abcactionnews.com/news/region-sarasota-manatee/cemetery-mystery-animals -trying-to-dig-up-fresh-bodies.

"Badgers dig up graves and leave human remains around cemetery, but protected animals cannot be removed." *Telegraph*, September 13, 2016. https://www.telegraph .co.uk/news/2016/09/13/badgers-dig-up-graves-and-leave-human-remains-around -cemetery-bu/.

Martin, Montgomery. *The History, Antiquities, Topography, and Statistics of Eastern India*, vol 2. London: William H. Allen, 1838.

What would happen if you swallowed a bag of popcorn before you died and were cremated?

Gale, Christopher P., and Graham P. Mulley. "Pacemaker explosions in crematoria: problems and possible solutions." *Journal of the Royal Society of Medicine* 95, no. 7 (July 2002). https://www.ncbi.nlm.nih.gov/pmc/articles/PMC1279940/.

Kinsey, Melissa Jayne. "Going Out With a Bang." *Slate*, October 26, 2017. http://www.slate.com/articles/technology/future_tense/2017/10/implanted_medical_devices_are_saving_lives_they_re_also_causing_exploding.html.

If someone is trying to sell a house, do they have to tell the buyer someone died there?

Adams, Tyler. "Is it required to disclose a murder on a property in Texas?" *Architect Tonic* (blog), December 22, 2010. https://tdatx.wordpress.com/2010/12/22/is-it-required-to-disclose-a-murder-on-a-property-in-texas/.

Griswold, Robert. "Death in a rental unit must be disclosed." *SFGate*, June 24, 2007. https://www.sfgate.com/realestate/article/Death-in-a-rental-unit-must-be-disclosed-2584502.php.

DiedInHouse website. https://www.diedinhouse.com/.

Bray, Ilona. "Selling My House: Do I Have to Disclose a Previous Death Here?" *Nolo*, n.d. https://www.nolo.com/legal-encyclopedia/selling-my-house-do-i-have-disclose-previous-death-here.html.

Spengler, Teo. "Do Apartments Have to Disclose if There's Been a Death?" *SFGate*, updated December 11, 2018. https://homeguides.sfgate.com/apartments-disclose-theres-death-44805.html.

Albrecht, Emily. "Dead Men Help No Sales." American Bar Association, n.d. https://www.americanbar.org/groups/young_lawyers/publications/tyl/topics/real-estate/dead-men-help-no-sales/.

"Do I have to Disclose a Death in the House?" Marcus Brown Properties, February 23, 2015. http://www.portlandonthemarket.com/blog/do-i-have-disclose-death-house/.

Order of the Good Death. "How Close Is Too Close? When Death Affects Real Estate." http://www.orderofthegooddeath.com/close-close-death-affects-real-estate.

White, Stephen Michael. "Should Landlords Tell Tenants About a Previous Death in the Property?" Rentprep, November 5, 2013. https://www.rentprep.com/leasing-questions/landlords-disclose-previous-death/.

Thompson, Jayne. "Does a Violent Death in a House Have to Be Disclosed?" *SFGate*, updated November 5, 2018. https://homeguides.sfgate.com/violent-death-house-disclosed-92401.html.

What if they make a mistake and bury me when I'm just in a coma?

"Have People Been Buried Alive?" *Snopes*. https://www.snopes.com/fact-check/just-dying-to-get-out/.

Valentine, Carla. "Why waking up in a morgue isn't quite as unusual as you'd think." *Guardian*, November 14, 2014. https://www.theguardian.com/commentisfree/2014/nov/14/waking-morgue-death-janina-kolkiewicz.

Olson, Leslie C. "How Brain Death Works." *How Stuff Works*. https://science.howstuffworks.com/life/inside-the-mind/human-brain/brain-death3.htm.

Senelick, Richard. "Nobody Declared Brain Dead Ever Wakes Up Feeling Pretty Good." *Atlantic*, February 27, 2012. https://www.theatlantic.com/health/archive/2012/02/nobody-declared-brain-dead-ever-wakes-up-feeling-pretty-good/253315/.

Brain Foundation. "Vegetative State (Unresponsive Wakefulness Syndrome)." http://brainfoundation.org.au/disorders/vegetative-state.

"Buried Alive: 5 Historical Accounts." *Innovative History*. http://innovativehistory.com/ih-blog/buried-alive.

Schoppert, Stephanie. "Back From the Dead: 8 Unbelievable Resurrections From History." *History Collection*. https://historycollection.co/back-dead-8-unbelievable-resurrections-throughout-history/.

"Beds, Herts & Bucks: Myths and Legends." BBC, November 10, 2014. http://www.bbc.co.uk/threecounties/content/articles/2008/09/29/old_mans_day_feature.shtml.

Adams, Susan. "A Fate Worse Than Death." *Forbes*, March 5, 2001. https://www.forbes.com/forbes/2001/0305/193.html#eb157542f39f.

Black Doctor. "Brain Dead vs. Coma vs. Vegetative State: What's the Difference?" https://blackdoctor.org/454040/brain-dead-vs-coma-vs-vegetative-state-whats-the-difference/.

Kiel, Carly. "12 Amazing Real-Life Resurrection Stories." *Weird History*. https://www.ranker.com/list/top-12-real-life-resurrection-stories/carly-kiel.

Marshall, Kelli. "4 People Who Were Buried Alive (And How They Got Out)." *Mental Floss*, February 15, 2014. http://mentalfloss.com/article/54818/4-people-who-were-buried-alive-and-how-they-got-out.

Lumen. "Lower-Level Structures of the Brain." https://courses.lumenlearning.com/teachereducationx92x1/chapter/lower-level-structures-of-the-brain/.

Morton, Ella. "Scratch Marks on Her Coffin: Tales of Premature Burial." *Slate*, October 7, 2014. https://slate.com/human-interest/2014/10/buried-alive-victorian-vivisepulture-safety-coffins-and-rufina-cambaceres.html.

Haynes, Sterling. "Special Feature: Tobacco Smoke Enemas." *BC Medical Journal* 54, no. 10 (December 2012): 496–97. https://www.bcmj.org/special-feature/special-feature-tobacco-smoke-enemas.

Icard, Severin. "The Written Test of the Dead and the Bump Map of Crime." *JF Ptak Science Books* (blog), post 2062. https://longstreet.typepad.com/ thesciencebookstore/2013/07/jf-ptak-science-books-post-2062-the-determination-of -the-occurrence-of-death-was-a-major-medical-feature-of-the-19th-centur.html.

Association of Organ Procurement Organizations. "Declaration of Brain Death." http://www.aopo.org/wikidonor/declaration-of-brain-death/.

What would happen if you died on a plane?

Clark, Andrew. "Airline's new fleet includes a cupboard for corpses." *Guardian*, May 10, 2004. https://www.theguardian.com/business/2004/may/11/ theairlineindustry.travelnews.

Do bodies in the cemetery make the water we drink taste bad?

Anderson, L. V. "Dead in the Water." *Slate*, February 22, 2013. http://www.slate .com/articles/health_and_science/explainer/2013/02/elisa_lam_corpse_water_ what_diseases_can_you_catch_from_water_that_s_touched.html.

Sack, R. B., and A. K. Siddique. "Corpses and the spread of cholera." *Lancet* 352, no. 9140 (November 14, 1998): 1570. https://www.ncbi.nlm.nih.gov/pubmed/9843100.

Oliveira, Bruna, Paula Quintero, Carla Caetano, Helena Nadais, Luis Arroja, Eduardo Ferreira da Silva, and Manuel Senos Matias. "Burial grounds' impact on groundwater and public health: an overview." *Water and Environment Journal* 27, no. 1 (March 1, 2013). https://www.deepdyve.com/lp/wiley/burial-grounds-impact -on-groundwater-and-public-health-an-overview-wquMEqoYLq?articleList=%2Fsea rch%3Fquery%3Dcorpse%2Bpreservation%26page%3D7.

Bourel, Benoit, Gilles Tournel, Valéry Hédouin, and Didier Gosset. "Entomo- fauna of buried bodies in northern France." *International Journal of Legal Medicine* 118, no. 4 (April 28, 2004). https://www.deepdyve.com/lp/springer-journals/ entomofauna-of-buried-bodies-in-northern-france-23c5gd95d0?articleList=%2Fsear ch%3Fquery%3Dcorpse%2Bpreservation%26page%3D10.

Bloudoff-Indelicato, Mollie. "Arsenic and Old Graves: Civil War-Era Cemeteries May Be Leaking Toxins." *Smithsonian*, October 30, 2015. https://www .smithsonianmag.com/science-nature/arsenic-and-old-graves-civil-war-era -cemeteries-may-be-leaking-toxins-180957115/.

I went to the show where dead bodies with no skin play soccer. Can we do that with my body?

Bodyworlds. "Body Donation." https://bodyworlds.com/plastination/bodydonation/.

Burns, L. "Gunther von Hagens' BODY WORLDS: selling beautiful education." *American Journal of Bioethics* 7, no. 4 (April 2007): 12–23. https://www.ncbi.nlm .nih.gov/pubmed/17454986.

Engber, Daniel. "The Plastinarium of Dr. Von Hagens." *Wired*, February 12, 2013. https://www.wired.com/2013/02/ff-the-plastinarium-of-dr-von-hagens/.

Ulaby, Neda. "Origins of Exhibited Cadavers Questioned." *All Things Considered*, NPR, August 11, 2006. https://www.npr.org/templates/story/story.php?storyId=5637687.

BODIES The Exhibition, "Bodies the Exhibition Disclaimer," http://www.premierexhibitions.com/exhibitions/4/4/bodies-exhibition/bodies-exhibition-disclaimer. Accessed April 1, 2019.

If someone is eating something when they die, does their body digest that food?

Bisker, C., and T. Komang Ralebitso-Senior. "Chapter 3—The Method Debate: A State-of-the-Art Analysis of PMI Investigation Techniques." *Forensic Ecogenomics* 2018: 61–86. https://doi.org/10.1016/b978-0-12-809360-3.00003-5.

Madea, B. "Methods for determining time of death." *Forensic Science, Medicine, and Pathology* 12, no. 4 (June 4, 2016): 451–485. https://doi.org/10.1007/s12024-016-9776-y.

WebMD. "Your Digestive System." https://www.webmd.com/heartburn-gerd/your-digestive-system#1.

Suzuki, Shigeru. "Experimental studies on the presumption of the time after food intake from stomach contents." *Forensic Science International* 35, nos. 2–3 (October–November 1987): 83–117. https://doi.org/10.1016/0379-0738(87)90045-4.

Can everybody fit in a casket? What if they're really tall?

Memorials.com. "Oversized Caskets." https://www.memorials.com/oversized-caskets.php.

Collins, Jeffrey. "Judge closes funeral home that cut off a man's legs." *Post and Courier*, July 14, 2009. https://www.postandcourier.com/news/judge-closes-funeral-home-that-cut-off-a-man-s/article_53334715-8122-510f-9945-dc84e1d3bf6f.html.

Fast Caskets. "What size casket do I need for my loved one?" https://blog.fastcaskets.com/2016/05/31/what-size-casket-do-i-need-for-my-loved-one/.

US Funerals Online. "Can an Obese Person be Cremated?" http://www.us-funerals.com/funeral-articles/can-an-obese-person-be-cremated.html#.W9y5P3pKjOQ.

Cremation Advisor. "What happens during the cremation process? From the Funeral Home receiving the deceased for cremation, to giving the family the cremated remains." DFS Memorials, July 26, 2018. http://dfsmemorials.com/cremation-blog/tag/oversize-cremation/.

US Cremation Equipment. "Products: Human Cremation Equipment." https://www.uscrematiomequipment.com/products/.

Can someone donate blood after they die?

Babapulle, C. J., and N. P. K. Jayasundera. "Cellular Changes and Time since Death." *Medicine, Science and the Law* 33, no. 3 (July 1, 1993): 213–22. https://doi .org/10.1177/002580249303300306.

Kevorkian, J., and G. W. Bylsma. "Transfusion of Postmortem Human Blood." *American Journal of Clinical Pathology* 35, no. 5 (May 1, 1961): 413–19. https://doi .org/10.1093/ajcp/35.5.413.

M. Sh. Khubutiya, S. A. Kabanova, P. M. Bogopol'skiy, S. P. Glyantsev, and V. A. Gulyaev. "Transfusion of cadaveric blood: an outstanding achievement of Russian transplantation, and transfusion medicine (to the 85th anniversary since the method establishment)." *Transplantologiya* 4 (2015): 61–73. https://www .jtransplantologiya.ru/jour/article/view/85?locale=en_US.

Moore, Charles L., John C. Pruitt, and Jesse H. Meredith. "Present Status of Cadaver Blood as Transfusion Medium: A Complete Bibliography on Studies of Postmortem Blood." *Archives of Surgery* 85, no. 3 (1962): 364–70. https:// jamanetwork.com/journals/jamasurgery/article-abstract/560305.

Roach, Mary. *Stiff: The Curious Lives of Human Cadavers.* New York and London: W. W. Norton, 2003. See pp. 228–32.

Vásquez-Valdés, E., A. Marín-López, C. Velasco, E. Herrera-Martínez, A. Pérez-Rojas, R. Ortega-Rocha, M. Aldama-Romano, J. Murray, and D. C. Barradas-Guevara. [Blood Transfusions from Cadavers]. Article in Spanish. *Revista de Investigación Clínica* 41, no. 1 (January–March 1989): 11-6. https://www.ncbi.nlm .nih.gov/pubmed/2727428.

Nebraska Department of Health and Human Services. "Organ, Eye and Tissue Donation." http://dhhs.ne.gov/publichealth/Pages/otd_index.aspx.

We eat dead chickens, why not dead people?

Price, Michael. "Why don't we eat each other for dinner? Too few calories, says new cannibalism study." *Science*, April 6, 2017. http://www.sciencemag.org/ news/2017/04/why-don-t-we-eat-each-other-dinner-too-few-calories-says-new -cannibalism-study.

Cole, James. "Assessing the Calorific Significance of Episodes of Human Cannibalism in the Palaeolithic." *Scientific Reports* 7, article no. 44707 (April 6, 2017). https://www.nature.com/articles/srep44707.

Liberski, Pawel P., Beata Sikorska, Shirley Lindenbaum, Lev G. Goldfarb, Catriona McLean, Johannes A. Hainfellner, and Paul Brown. "Kuru: Genes, Cannibals and Neuropathology." *Journal of Neuropathology and Experimental Neurology* 71, no. 2 (February 2012). https://www.ncbi.nlm.nih.gov/pmc/articles/PMC5120877/.

González Romero, María Soledad, and Shira Polan. "Cannibalism Used to Be a Popular Medical Remedy—Here's Why Humans Don't Eat Each Other

Today." *Business Insider*, June 7, 2018. https://www.businessinsider.com/why-self
-cannibalism-is-bad-idea-2018-5.

Wordsworth, Rich. "What's wrong with eating people?" *Wired*, October 28, 2017.
https://www.wired.co.uk/article/lab-grown-human-meat-cannibalism.

Borreli, Lizette. "Side Effects Of Eating Human Flesh: Cannibalism Increases Risk
of Prion Disease, And Eventually Death." *Medical Daily*, May 19, 2017. https://www
.medicaldaily.com/side-effects-eating-human-flesh-cannibalism-increases-risk-prion
-disease-and-417622.

Scutti, Susan. "Eating Human Brains Led To A Tribe Developing Brain
Disease-Resistant Genes." *Medical Daily*, June 11, 2015. https://www
.medicaldaily.com/eating-human-brains-led-tribe-developing-brain-disease
-resistant-genes-337672.

Rettner, Rachael. "Eating Brains: Cannibal Tribe Evolved Resistance to Fatal Disease." *Live Science*, June 12, 2015. https://www.livescience.com/51191-cannibalism
-prions-brain-disease.html.

Rense, Sarah. "Let's Talk About Eating Human Meat." *Esquire*, April 7, 2017.
https://www.esquire.com/lifestyle/health/news/a54374/human-body-parts-calories/.

"Table 1: Average weight and calorific values for parts of the human body." *Scientific Reports*. https://www.nature.com/articles/srep44707/tables/1.

Katz, Brigit. "New Study Fleshes Out the Nutritional Value of Human Meat."
Smithsonian, April 7, 2017. https://www.smithsonianmag.com/smart-news/ancient
-cannibals-did-not-eat-humans-nutrition-study-says-180962823/.

What happens when a cemetery is full of bodies and you can't add any more?

Biegelsen, Amy. "America's Looming Burial Crisis." *CityLab*, October 31, 2012.
https://www.citylab.com/equity/2012/10/americas-looming-burial-crisis/3752/.

Wallis, Lynley, Alice Gorman, and Heather Burke. "Losing the plot: death is
permanent, but your grave isn't." *The Conversation*, November 5, 2014. http://
theconversation.com/losing-the-plot-death-is-permanent-but-your-grave-isnt-33459.

National Center for Health Statistics. "Deaths and Mortality." Centers for Disease
Control and Prevention, updated May 3, 2017. https://www.cdc.gov/nchs/fastats/
deaths.htm.

de Sousa, Ana Naomi. "Death in the city: what happens when all our cemeteries
are full?" *Guardian*, January 21, 2015. https://www.theguardian.com/cities/2015/
jan/21/death-in-the-city-what-happens-cemeteries-full-cost-dying.

Ryan, Kate, and Christine Steinmetz. "Housing the dead: what happens when a
city runs out of space?" *The Conversation*, January 4, 2017. https://theconversation
.com/housing-the-dead-what-happens-when-a-city-runs-out-of-space-70121.

National Environmental Agency, Singapore. "Post Death Matters." Updated June 20, 2018. https://www.nea.gov.sg/our-services/after-death/post-death-matters/burial -cremation-and-ash-storage.

Is it true people see a white light as they're dying?

Konopka, Lukas M. "Near death experience: neuroscience perspective." *Croatian Medical Journal* 56, no. 4 (August 2015): 392–93. https://doi.org/10.3325/cmj.2015 .56.392.

Mobbs, Dean, and Caroline Watt. "There is nothing paranormal about near-death experiences: how neuroscience can explain seeing bright lights, meeting the dead, or being convinced you are one of them." *Trends in Cognitive Sciences* 15, no. 10 (October 1, 2011): 447–49. https://doi.org/10.1016/j.tics.2011.07.010.

Lambert, E. H., and E. H. Wood. "Direct determination of man's blood pressure on the human centrifuge during positive acceleration." *Federation Proceedings* 5, no. 1 pt. 2 (1946): 59. https://www.ncbi.nlm.nih.gov/pubmed/21066321.

Owens, J. E., E. W. Cook, and I. Stevenson. "Features of 'near-death experience' in relation to whether or not patients were near death." *Lancet* 336, no. 8724 (November 10, 1990): 1175–77. https://www.ncbi.nlm.nih.gov/pubmed/1978037.

van Lommel, P., R. van Wees, V. Meyers, and I. Elfferich. "Near-death experience in survivors of cardiac arrest: a prospective study in the Netherlands." *Lancet* 358, no. 9298 (December 15, 2001): 2039–45. https://www.ncbi.nlm.nih .gov/pubmed/?term=Elfferich%20I%5BAuthor%5D&cauthor=true&cauthor_ uid=11755611.

Tsakiris, Alex. "What makes near-death experiences similar across cultures? L-O-V-E." *Skeptiko*, January 27, 2019. https://skeptiko.com/265-dr-gregory-shushan -cross-cultural-comparison-near-death-experiences/.

Why don't bugs eat people's bones?

Bloudoff-Indelicato, Mollie. "Flesh-Eating Beetles Explained." *National Geographic*, 17 January 17, 2013. https://blog.nationalgeographic.org/2013/01/17/flesh-eating -beetles-explained/.

Hall, E. Raymond, and Ward C. Russell. "Dermestid Beetles as an Aid in Cleaning Bones." *Journal of Mammalogy* 14, no. 4 (November 13, 1933): 372–74. https://doi .org/10.1093/jmammal/14.4.372.

Henley, Jon. "Lords of the flies: the insect detectives." *Guardian*, September 23, 2010. https://www.theguardian.com/science/2010/sep/23/flies-murder-natural -history-museum.

Monaco, Emily. "In 1590, Starving Parisians Ground Human Bones Into Bread." *Atlas Obscura*, October 29, 2018. https://www.atlasobscura.com/articles/what -people-eat-during-siege.

Vrijenhoek, Robert C., Shannon B. Johnson, and Greg W. Rouse. "A remarkable diversity of bone-eating worms (Osedax; Siboglinidae; Annelida)." *BMC Biology* 7 (November 2009): 74. https://doi.org/ 10.1186/1741-7007-7-74.

Zanetti, Noelia I., Elena C. Visciarelli, and Néstor D. Centeno. "Trophic roles of scavenger beetles in relation to decomposition stages and seasons." *Revista Brasileira de Entomologia* 59, no. 2 (2015): 132–37. http://dx.doi.org/10.1016/j.rbe.2015.03.009.

What happens when you want to bury someone but the ground is too frozen?

Liquori, Donna. "Where Death Comes in Winter, and Burial in the Spring." *New York Times*, May 1, 2005. https://www.nytimes.com/2005/05/01/nyregion/where -death-comes-in-winter-and-burial-in-the-spring.html.

Rylands, Traci. "The Frozen Chosen: Winter Grave Digging Meets Modern Times." *Adventures in Cemetery Hopping* (blog), February 27, 2015. https:// adventuresincemeteryhopping.com/2015/02/27/frozen-funerals-how-grave-digging -meets-modern-times/.

"Cold Winters Create Special Challenges for Cemeteries." *The Funeral Law Blog*, April 26, 2014. https://funerallaw.typepad.com/blog/2014/04/cold-winters-create -special-challenges-for-cemeteries.html.

Schworm, Peter. "Icy weather making burials difficult." Boston.com (website of *Boston Globe*), February 9, 2011. http://archive.boston.com/news/local/ massachusetts/articles/2011/02/09/icy_weather_making_burials_difficult/.

Lacy, Robyn. "Winter Corpses: What to do with Dead Bodies in colonial Canada." *Spade and the Grave* (blog), February 18, 2018. https://spadeandthegrave .com/2018/02/18/winter-corpses-what-to-do-with-dead-bodies-in-colonial-canada/.

"Funeral Planning: Winter Burials." iMortuary, blog post, November 2, 2013. https://www.imortuary.com/blog/funeral-planning-winter-burials/.

Rutledge, Mike. "Local woman hopes to restore historic vault at Hamilton cemetery." *Journal–News*, August 26, 2017. https://www.journal-news .com/news/local-woman-hopes-restore-historic-vault-hamilton-cemetery/ zUekzY68vA9biv8NVfqJVN/.

Can you describe the smell of a dead body?

Costandi, Moheb. "The smell of death." *Mosaic*, May 4, 2015. https://mosaicscience .com/extra/smell-death/.

Verheggen, François, Katelynn A. Perrault, Rudy Caparros Megido, Lena M. Dubois, Frédéric Francis, Eric Haubruge, Shari L. Forbes, Jean-François Focant, and Pierre-Hugues Stefanuto. "The Odor of Death: An Overview of Current Knowledge on Characterization and Applications." *BioScience* 67, no. 7 (July 1, 2017): 600–13. https://doi.org/10.1093/biosci/bix046.

Ginnivan, Leah. "The Dirty History of Doctors' Hands." *Method*, n.d. http://www
.methodquarterly.com/2014/11/handwashing/.

Haven, K. F. *100 Greatest Science Inventions of All Time*. Westport, CT: Libraries
Unlimited, 2005. See pp. 118–19.

Izquierdo, Cristina, José C. Gómez-Tamayo, Jean-Christophe Nebel, Leonardo
Pardo, and Angel Gonzalez. "Identifying human diamine sensors for death related
putrescine and cadaverine molecules." *PLoS Computational Biology* 14, no. 1 (January 11, 2018): e1005945. https://doi.org/10.1371/journal.pcbi.1005945.

What happens to soldiers who die far away in battle, or whose bodies are never found?

Kuz, Martin. "Death Shapes Life for Teams that Prepare Bodies of Fallen Troops
for Final Flight Home." *Stars and Stripes*, February 17, 2014. https://www.stripes
.com/death-shapes-life-for-teams-that-prepare-bodies-of-fallen-troops-for-final-flight
-home-1.267704.

Collier, Martin, and Bill Marriott. *Colonisation and Conflict 1750–1990*. London:
Heinemann, 2002.

Beatty, William. *The Death of Lord Nelson*. London: T. Cadell and W. Davies, 1807.

Lindsay, Drew. "Rest in Peace? Bringing Home U.S. War Dead." *MHQ Magazine*,
Winter 2013. https://www.historynet.com/rest-in-peace-bringing-home-u-s-war
-dead.htm.

Quackenbush, Casey. "Here's How Hard It Is to Bring Home Remains of U.S. Soldiers, According to Experts." *Time*, July 27, 2018. http://time.com/5322001/north
-korea-war-remains-dpaa/.

Defense POW/MIA Accounting Agency. "Fact Sheets." http://www.dpaa.mil/
Resources/Fact-Sheets/.

Dao, James. "Last Inspection: Precise Ritual of Dressing Nation's War Dead." *New
York Times*, May 25, 2013. https://www.nytimes.com/2013/05/26/us/intricate-rituals
-for-fallen-americans-troops.html.

Can I be buried in the same grave as my hamster?

King, Barbara J. "When 'Whole-Family' Cemeteries Include Our Pets." NPR, May
18, 2017. https://www.npr.org/sections/13.7/2017/05/18/528736490/when-whole
-family-cemeteries-include-our-pets.

Green Pet-Burial Society. "Whole-Family Cemetery Directory – USA." https://
greenpetburial.org/providers/whole-family-cemeteries/.

Nir, Sarah Maslin. "New York Burial Plots Will Now Allow Four-Legged Companions." *New York Times*, October 6, 2016. https://www.nytimes.com/2016/10/07/
nyregion/new-york-burial-plots-will-now-allow-four-legged-companions.html.

Banks, T. J. "Why Some People Want to Be Buried With Their Pets." *Petful*, August 28, 2017. https://www.petful.com/animal-welfare/can-pet-buried/.

Vatomsky, Sonya. "The Movement to Bury Pets Alongside People." *Atlantic*, October 10, 2017. https://www.theatlantic.com/family/archive/2017/10/whole-family -cemeteries/542493/.

Blain, Glenn. "New Yorkers can be buried with their pets under new law." *New York Daily News*, September 26, 2016. https://www.nydailynews.com/new-york/new -yorkers-buried-pets-new-law-article-1.2807109.

LegalMatch. "Pet Burial Laws." https://www.legalmatch.com/law-library/article/pet -burial-laws.html.

Isaacs, Florence. "Can You Bury Your Pet With You After You Die?" Legacy.com, "2 years ago" (from February 13, 2019). http://www.legacy.com/news/advice-and -support/article/can-you-bury-your-pet-with-you-after-you-die.

Pruitt, Sarah. "Scientists Reveal Inside Story of Ancient Egyptian Animal Mummies." *History*, May 12, 2015. https://www.history.com/news/scientists-reveal-inside -story-of-ancient-egyptian-animal-mummies.

Faaberg, Judy. "Washington state seeks to force cemeteries to bury pets with their humans." International Cemetery, Cremation and Funeral Association, blog post, January 16, 2009. https://web.archive.org/web/20100215045254/http:/iccfa.com/ blogs/judyfaaberg/2009/01/15/washington-state-seeks-force-cemeteries-bury-pets -their-humans.

"Benji I." Find A Grave. https://www.findagrave.com/memorial/7376655/benji_i.

Street, Martin, Hannes Napierala, and Luc Janssens. "The late Paleolithic dog from Bonn-Oberkassel in context." In *The Late Glacial Burial from Oberkassel Revisited*, edited by L. Giemsch and R. W. Schmitz. *Rheinische Ausgrabungen* 72: 253–74. https://www.researchgate.net/publication/284720121_Street_M_ Napierala_H_Janssens_L_2015_The_late_Palaeolithic_dog_from_Bonn -Oberkassel_in_context_In_The_Late_Glacial_Burial_from_Oberkassel_ Revisited_L_Giemsch_R_W_Schmitz_eds_Rheinische_Ausgrabungen_72.

Will my hair keep growing in my coffin after I'm buried?

Palermo, Elizabeth. "30-Foot Fingernails: The Curious Science of World's Longest Nails." *Live Science*, October 1, 2015. https://www.livescience.com/52356-science-of -worlds-longest-fingernails.html.

Hammond, Claudia. "Do your hair and fingernails grow after death?" *BBC Future*, May 28, 2013. http://www.bbc.com/future/story/20130526-do-your-nails-grow-after -death.

Aristotle. "De Generatione Animalium." *The Works of Aristotle*, edited by J. A. Smith and W. D. Ross, vol. 5. Oxford: Clarendon Press, 1912.

"Editorial: The Druce Case." *Edinburgh Medical Journal* 23: 97–100. Edinburgh and London: Young J. Pentland, 1908.

Can I use human bones from a cremation as jewelry?

Nora Menkin, Executive Director at People's Memorial Association and the Co-op Funeral Home, was an important source for this section.

Kim, Michelle. "How Cremation Works." *How Stuff Works.* https://science.howstuffworks.com/cremation2.htm.

FuneralWise. "The Cremation Process." https://www.funeralwise.com/plan/cremation/cremation-process/.

Chesler, Caren. "Burning Out: What Really Happens Inside a Crematorium." *Popular Mechanics*, March 1, 2018. https://www.popularmechanics.com/science/health/a18923323/cremation/.

Absolonova, Karolina, Miluše Dobisíková, Michal Beran, Jarmila Zoková, and Petr Veleminsky. "The temperature of cremation and its effect on the microstructure of the human rib compact bone." *Anthropologischer Anzeiger* 69, no. 4 (November 2012): 439–60. https://www.researchgate.net/publication/235364719_The_temperature_of_cremation_and_its_effect_on_the_microstructure_of_the_human_rib_compact_bone.

The Funeral Source. "Asian Funeral Traditions." http://thefuneralsource.org/trad140205.html.

Treasured Memories. "Japanese Cremation Ceremony: A Celebration of Life." https://tmkeepsake.com/blog/celebration-life-japenese-cremation-ceremony/.

Perez, Ai Faithy. "The Complicated Rituals of Japanese Funerals." *Savvy Tokyo*, October 21, 2015. https://savvytokyo.com/the-complicated-rituals-of-japanese-funerals/.

LeBoutillier, Linda. "Memories of Japan: Cemeteries and Funeral Customs." *Random Thoughts . . . a beginner's blog*, January 8, 2014. http://mettahu.blogspot.com/2014/01/memories-of-japan-cemeteries-and.html.

Imaizumi, Kazuhiko. "Forensic investigation of burnt human remains." *Research and Reports in Forensic Medical Science* 2015, no. 5 (December 2015): 67–74. https://www.dovepress.com/forensic-investigation-of-burnt-human-remains-peer-reviewed-fulltext-article-RRFMS.

North Carolina Legislature. "Article 13F: Cremations." https://www.ncleg.net/EnactedLegislation/Statutes/PDF/ByArticle/Chapter_90/Article_13F.pdf.

Did mummies stink when they were wrapped?

"The Chemistry of Mummification." *Compound Interest*, October 27, 2016. http://www.compoundchem.com/2016/10/27/mummification/.

Krajick, Kevin. "The Mummy Doctor." *New Yorker*, May 16, 2005.

Smithsonian Institution. "Ancient Egypt/ Egyptian Mummies." https://www.si.edu/spotlight/ancient-egypt/mummies.

At my grandma's wake she was wrapped in plastic under her blouse. Why would they do that?

Faull, Christina, and Kerry Blankley. "Table 7.2: Care for a Patient After Death." *Palliative Care*. 2nd edition. Oxford, UK: Oxford University Press, 2015.

Smith, Matt. "Embalming the Severe EDEMA Case: Part 1." *Funeral Business Advisor*, January 26, 2016. https://funeralbusinessadvisor.com/embalming-the-severe-edema-case-part-1/funeral-business-advisor.

Payne, Barbara. "Winter 2015 dodge magazine." https://issuu.com/ddawebdesign/docs/winter_2015_dodge_magazine.